Research in Biotechnology

Research in Biotechnology

P. R. Gilbert

ANMOL PUBLICATIONS PVT. LTD.
NEW DELHI - 110 002 (INDIA)

ANMOL PUBLICATIONS PVT. LTD.
H.O.: 4374/4B, Ansari Road, Darya Ganj
New Delhi-110 002 (India)
Ph.: 23278000, 23261597
B.O.: 1015, Ist Main Road, BSK IIIrd Stage
IIIrd Phase, IIIrd Block
Bangalore - 560 085 (Karnataka)
Ph: 080-41723429, Tel/Fax: 080-2672 3604
Visit us at: www.anmolpublications.com

Research in Biotechnology

First Published, 2008

ISBN 978-81-261-3270-6

PRINTED IN INDIA

Printed at Mehra Offset Press, Delhi.

Contents

Preface

This book entitled "Research in Biotechnology" is written with an aim to present readers with the new researches in biotechnology in a user-friendly manner, including the role of select key visionaries and personalities in modern global biotechnology sector.

Advances in science and technology have allowed researchers to improve enzymes through modifications to enzyme producing microorganisms, or through direct changes to the enzymes themselves. By using recombinant DNA technology, microorganisms may be genetically modified to produce a desired enzyme under specific conditions. This is accomplished through recombinant DNA technology, whereby small circular pieces of DNA, known as plasmids, are used to insert enzyme-producing genes into the genomes of organisms that possess another desirable trait, such as the ability to thrive on inexpensive nutrients. Therefore, both the enzyme and the original trait will be expressed in a single recombinant microorganism. All enzymes are made of proteins, which are large molecules formed from basic units, called amino acids, strung together like beads on a chain. To form functional enzymes, long chains of amino acids must be folded properly. Some enzymes consist of only one chain, whereas others are made of several chains that fit together. Scientists are using technology to study how proteins are formed, how they fold, and how they function. By studying the relation between the structure of a protein and how it functions, they are developing ways to improve and engineer enzymes. Changing one or more amino acids, and/or changing the way the amino acid chains fold and fit together may modify proteins. Foodstuffs made of genetically modified crops that are currently available (mainly maize, soybean, and oilseed rape) have been judged safe to eat, and the methods used to test them have been deemed appropriate. These conclusions represent the consensus of the scientific evidence surveyed by the International Council for Science (ICSU) and are consistent with the views of the World Health Organization (WHO). However, the lack of evidence of negative effects does not mean that new genetically modified foods are without risk. The possibility of long-term effects from genetically modified plants cannot be excluded and must be examined on a case-by-case basis. New techniques are being developed to address concerns, such as the possibility of the unintended transfer of antibiotic-resistance genes. Genetic engineering of plants could also offer some direct and indirect health benefits to consumers, for instance by improving nutritional quality or reducing pesticide use. Scientists recommend that food safety assessment should take place on a case-by-case basis before genetically modified food is brought to the market. In such assessments, foodstuffs derived from genetically modified plants are compared to their conventional counterparts, which are generally considered safe due to their long history of use. This comparison considers to what extent different foodstuffs can cause harmful effects or allergies and how much nutrients they contain.

"Transgenic organism" is now the preferred term for genetically modified organisms with extra-genome (foreign genetic) information, as opposed to "genetically engineered" or "genetically modified" organisms (which may refer to changes made within the genome such as amplification or deletion of genes). The first Genetically Engineered drug was human insulin approved by the USA's FDA in 1982. Another early application of GE was to create human growth hormone as replacement for a drug that was previously extracted from human cadavers. In 1986 the FDA approved the first genetically engineered vaccine for humans, for hepatitis B. Since these early uses of the technology in medicine the use of the GE has expanded to supply many drugs and vaccines. One of the best-known applications of genetic engineering is that of the creation of genetically modified organisms (GMOs). There are potentially momentous biotechnological applications of GM, for example oral vaccines produced naturally in fruit, at very low cost. A radical ambition of some groups is human enhancement via genetics, eventually by molecular engineering. DNA sequencing is a technique, which is used to identify each base in DNA. Although the costs of DNA sequencing has dropped dramatically, the NIH estimates it costs at least $10 million to sequence 3 billion base pairs—the size of the whole human genome. Although there has been a tremendous revolution in the biological sciences in the past twenty years, there is still a great deal that remains to be discovered. The completion of the sequencing of the human genome, as well as the genomes of most agriculturally and scientifically important plants and animals, has increased the possibilities of genetic research immeasurably.

Expedient and inexpensive access to comprehensive genetic data has become a reality, with billions of sequenced nucleotides already online and annotated. Now that the rapid sequencing of arbitrarily large genomes has become a simple, if not trivial affair, a much greater challenge will be elucidating function of the extraordinarily complex web of interacting proteins, dubbed the proteome, that constitutes and powers all living things. Genetic engineering has become the gold standard in protein research, and major research progress has been made using a wide variety of techniques. Research in biochemicals continues to grow rapidly.

Many large multinational companies are investing time and money into producing chemicals that can be economically viable and environmentally friendly. For example, many chemical companies are teaming up with biotechnology companies to develop products based on biochemicals. Cargill Dow currently makes polylactic acid from corn kernels but has plans to switch to cheaper feedstocks, such as corn stalks, wheat straw, rice hulls, and sawdust, so that biochemicals can be more competitive in the marketplace. Current research is also focused on developing genetically modified microorganisms for use in specific chemical productions. Pollution due to the use of petroleum-based products may be eliminated by the use of biochemicals. Whereas petroleum-based products do not biodegrade and accumulate in waste landfills, biochemicals degrade completely and do not contribute to waste problems. As well, biochemicals are made from biomass; therefore, their production and use are sustainable.

This book consists of—Biotechnology Research: National Achievement Survey; Policy for Research in Biotechnology: A Model; Mutant Germplasm Characterization: A Global Perspective; and Stem Cell Research in USA: A Low and Policy Perspective etc. besides a large body of list of acronyms, glossary of relevant terms, extensive bibliography, list of websites and links, for further referencing and research.

—Editor

List of Acronyms

ABRF	—	Association of Biomolecular Resource Facilities
ACMG	—	American College of Medical Genet.
ACS	—	American Chemical Society
ADA	—	Americans with Disabilities Act
AEC	—	Atomic Energy Commission
AES	—	American Electrophoresis Society
AFMR	—	American Federation of Medical Research
AFIP/ARP	—	Armed Forces Inst. of Pathol./Am. Registry of Pathol.
AGSG	—	Alliance of Genetic Support Groups (now Genetic Alliance)
AGT	—	Association of Genetic Technologists
AHA	—	American Heart Association
AIBS	—	Am. Inst. of Biol. Soc.
AMIA	—	American Medical Informatics Association
BTCI	—	BioPharmaceutical Technology Center Institute
CF	—	Cystic Fibrosis
CFF	—	Cystic Fibrosis Foundation
CHH	—	Cartilage-hair Hypoplasia

CHI	—	Cambridge Healthtech Inst.
CHOP	—	Children's Hospital of Philadelphia
CIMB	—	Center for International Meeting on Biology
CIOMS	—	Council for International Organizations of Medical Sciences
CIRB	—	Colorado Institute for Research in Biotechnology
CLMA	—	Clinical Laboratory Management Association
CMT	—	Charcot-Marie-Tooth
EMBO	—	European Molecular Biology Organisation
EMG	—	Encyclopedia of the Mouse Genome
EMS	—	Environmental Mutagen Society
EORTC	—	European Organization for Research and Treatment of Cancer
ERDA	—	Energy Research and Development Administration
ERI	—	Eleanor Roosevelt Institute
ES	—	Embryonic Stem
GBASE	—	Genome Database of the Mouse
GBR	—	Global Business Research
GDB	—	Genome Database
GMD	—	Genomic Map Design
GPI	—	Genetics and Public Issues Program
GRAIL	—	Gene Recognition and Analysis Internet Link
GRC	—	Gordon Res. Conf.

HMDP	—	Homology Database
ICGEB	—	International Centre for Genetic Engineering and Biotechnology
ICHG	—	International Congress of Human Genetics
ICPEMC	—	International Commission on Protection Against Environmental Mutagens and Carcinogens
ICRF	—	Imperial Cancer Research Fund
IEEE	—	Institute of Electrical and Electronics Engineers
IG	—	IntelliGenetics
IGES	—	International Genetics Epidemiology Societies
IJCAI	—	International Joint Conference on Artificial Intelligence
IMA	—	Institute for Mathematics and its Applications
MBC	—	Massachusetts Biotechnology Council
MBL	—	Marine Biological Laboratory
MOD	—	March of Dimes
MOU	—	Memorandum of Understanding
MRC	—	Medical Research Council
NAPBC	—	Natl. Action Plan on Breast Cancer
NAS	—	Natl. Academy of Sciences
NASA	—	Natl. Aeronautics and Space Administration
NCGR	—	National Center for Genome Resources

NCI	—	Natl. Cancer Institute
NCSA	—	National Center for Supercomputing Applications
NCSL	—	National Conference of State Legislatures
NCSU	—	North Carolina State University
NEH	—	National Endowment for the Humanities
NFCR	—	National Foundation for Cancer Research
NFID	—	National Foundation for Infectious Diseases
NHGRI	—	Natl. Human Genome Res. Inst.
NICHD	—	National Institute of Child Health and Human Development
NIGMS	—	National Institute of General Medical Sciences
NIH	—	National Institutes of Health
NIGMS	—	Natl. Institute of General Medical Sciences
NIST	—	Natl. Inst. of Standards and Technology
NLGLP	—	National Laboratory Gene Library Project
NLM	—	National Library of Medicine
NMHCC	—	National Managed Health Care Congress
NORD	—	National Organization for Rare Disorders
NRC	—	National Research Council

NRSA	—	National Research Service Award
NSF	—	Natl. Sci. Foundation
NSGC	—	National Society of Genetic Counselors
OHER	—	Office of Health and Environmental Research
OMIM	—	Online *Mendelian Inheritance in Man*
OPRR	—	Office of Protection from Res. Risks
ORAU	—	Oak Ridge Associated Universities
ORISE	—	Oak Ridge Inst. for Science and Education
ORF	—	open reading frame
ORNL	—	Oak Ridge National Laboratory
OSU	—	Oregon State University
PCR	—	Polymerase Chain Reaction
PFGE	—	Pulsed-field gel electrophresis
PG	—	Plant Genome
PIR	—	Protein Info. Resource
PNNL	—	Pacific Northwest National Laboratory
PRIM&R	—	Public Responsibility in Medicine & Research
RAPD	—	Random Amplified Polymorphic DNA
RARA	—	Retinoic Acid Receptor
RECOMB	—	Conference on Computational Molecular Biology
RFLP	—	Restriction Fragment Length Polymorphism

SCAN	—	Sequence Comparison ANalysis Program
SCE	—	School of Continuing Education
SCI	—	Society of Chemical Industry
SCW	—	Single-chromosome Workshop
SDC	—	San Diego Conference
SERGG	—	Southeast Regional Genetics Group
SHOM	—	Sequencing by Hybridization on Matrices
SIM	—	Society for Industrial Microbiology
SIMS	—	Societal Institute of the Mathematical Sciences
SIVB	—	Society for In Vitro Biology
SNL	—	Sandia National Laboratory
SNP	—	Single Nucleotide Polymorphism
SQL	—	Standard Query Language
UCLA	—	University of California, Los Angeles
UICM	—	Univ. of Illinois College of Medicine
WHS	—	Wolf-Hirschhorn syndrome
WLMG	—	Wellcome Laboratory for Molecular Genetics
WSES	—	World Scientific and Engineering Society

List of Glossary of Terms

Antigenic switching: The altering of a microorganism's surface antigens through genetic rearrangement, to elude detection by the host's immune system.

Bacteriorhodopsin: Pigmented protein found in the plasma membrane of a salt-loving bacterium, *Halobacterium halobium*; it pumps protons out of the cell in response to light.

Basal lamina (plural basal laminae): Thin mat of extracellular matrix that separates epithelial sheets, and many types of cells such as muscle cells or fat cells, from connective tissue. Sometimes called a basement membrane.

Beta-DNA: The normal form of DNA found in biological systems, which exists as a right-handed helix.

Biomotors: Driven by energy sources such as *adenosine triphosphate (ATP)* for chemical transduction and other processes. These biomotors are considered to be biomolecular and are discussed in the body of this report, but strictly speaking they do not conform to the panel's definition of self- assembly.

Biosensor: The term "biosensor" is a general designation that denotes either a sensor to detect a biological substance or a sensor which incorporates the use of biological molecules such as antibodies or enzymes. Biosensors are a subcategory of chemical sensors.

Chloramphenicol: An antibiotic that interferes with protein synthesis.

Chromatid: Each of the two daughter strands of a duplicated chromosome joined at the centromere during mitosis and meiosis.

Cistron: A DNA sequence that codes for a specific polypeptide; a gene. See DNA, Gene.

Clone: An exact genetic replica of a specific gene or an entire organism. See Cloning.

Cloning vector: A DNA molecule capable of autonomous replication within the cloning host cell (e.g *E. coli*). The vectors contain restriction enzyme sites for insertion of foreign DNA. Cloning vectors are derived from bacterial plasmids, bacteriophages, or viruses.

Competency: An ephemeral state, induced by treatment with cold cations, during which bacterial cells are capable of uptaking foreign DNA.

Concatemer: A DNA segment composed of repeated sequences linked end to end.

Conjugation: The joining of two bacteria cells when genetic material is transferred from one bacterium to another.

Coulomb [C]: Measure of electrical charge: 1 C is an amount of charge equal to that of about 6.24x1018 electrons.

Coupling: A connection between more than one object or energy pathway so that together they function as a single unit.

Crossing-over: The exchange of DNA sequences between chromatids of homologous chromosomes during meiosis.

Cross-pollination: Fertilization of a plant from a plant with a different genetic makeup.

Crosstalk: Electromagnetic noise transmitted between leads or circuits in close proximity to each other.

Enabling science and technologies: Areas of research relevant to a particular goal, such as nanotechnology. Also, technology that "enables" other technology to advance, such as the transistor enabled the computer chip revolution, as did photolithography.

Encapsidation: Process by which a virus' nucleic acid is enclosed in a capsid. See Coat protein.

Endocrine cell: Specialized animal cell that secretes a hormone into the blood; usually part of a gland, such as the thyroid or pituitary gland.

Endocytosis: Uptake of material into a cell by an invagination of the plasma membrane and its internalization in a membrane-bounded vesicle.

Endophyte: An organism that lives inside another.

Endoplasmic reticulum (ER): Labyrinthine, membrane-bounded compartment in the cytoplasm of eucaryotic cells, where lipids are synthesized and membrane-bound proteins are made.

Energy [J]: Capacity for performing work or to cause heat flow. Like work itself, it is measured in Joules.

Environmental Protection Agency (EPA): The U.S. regulatory agency for biotechnology of microbes. The major laws under which the agency has regulatory powers are the Federal Insecticide, Fungicide, and Rodenticide Act (FIFRA); and the Toxic Substances Control Act (TSCA).

Enzymes: Proteins that control the various steps in all chemical reactions.

Ethidium bromide: A fluorescent dye used to stain DNA and RNA. The dye fluoresces when exposed to UV light.

Eukaryote: An organism whose cells possess a nucleus and other membrane-bound vesicles, including all members of the protist, fungi, plant and animal kingdoms; and excluding viruses, bacteria, and blue-green algae. See Prokaryote.

Eurisko: A computer program developed by Professor Douglas Lenat which is able to apply heuristic rules for performing various tasks, including the invention of new heuristic rules.

FIFRA: The Federal Insecticide, Fungicide, and Rodenticide Act. See Environmental Protection Agency.

Fingerprint: A set of molecular markers sufficiently diverse to identify particular individuals with reasonable certainty.

Flanking region: The DNA sequences extending on either side of a specific locus or gene.

Food and Drug Administration (FDA): The U.S. agency responsible for regulation of biotechnology food products. The major laws under which the agency has regulatory powers include the Food, Drug, and Cosmetic Act; and the Public Health Service Act.

Gel, agarose: A substance used to separate DNA or RNA fragments by size. Used for southern and northern blots.

Gel, polyacrylamide: A substance used to separate proteins by size. Used for western blots.

GEM: A genetically engineered microorganism.

Gene amplification: The presence of multiple genes. Amplification is one mechanism through which proto-oncogenes are activated in malignant cells.

Gene cloning: The process of synthesizing multiple copies of a particular DNA sequence using a bacteria cell or another organism as a host. See DNA, Host.

Gene expression: The process of producing a protein from its DNA- and mRNA-coding sequences.

Gene flow: The exchange of genes between different but (usually) related populations.

Gene gun: Transformation technique that uses accelerated particles coated with DNA to introduce foreign DNA into recipient plant.

Gene insertion: The addition of one or more copies of a normal gene into a defective chromosome.

Gene linkage: The hereditary association of genes located on the same chromosome.

Gene modification: The chemical repair of a gene's defective DNA sequence. See DNA.

Gene pool: The totality of all alleles of all genes of all individuals in a particular population.

Genetic drift: Random variation in gene frequency from one generation to another.

Genetic engineering: Genetic engineering is the process of manually adding new DNA on a molecular level with the goal of adding one or more new traits that are not already found in that organism. For the purposes of this web site, genetic engineering includes recombinant DNA and gene splicing technologies.

Genetic marker: A gene or group of genes used to "mark" or track the action of microbes.

Genome: The genetic complement contained in the chromosomes of a given organism, usually the haploid chromosome state. The full chromosome set containing all the genes of a particular individual.

Genomics:: The study of the structure and function of genomes. Genomics usually involves high speed sequencing of the DNA and computer searches for sequences that code for genes.

Genotype: The sum total of the genetic information of an organism including the linkage relationships between genes. The genotype, modified by environment, determines the phenotype.

Genus: A category including closely related species. Interbreeding between organisms within the same category can occur.

Germ cell (germ line) gene therapy: The repair or replacement of a defective gene within the gamete-forming tissues, which produces a heritable change in an organism's genetic constitution.

Green revolution: Advances in genetics, petrochemicals, and machinery that culminated in a dramatic increase in crop productivity during the third quarter of the 20th century.

GUI: Graphical User Interface — hardware, software, and firmware that produces the display on modern personal computers.

Heterochromatin: Dark-stained regions of chromosomes thought to be for the most part genetically inactive.

Heteroduplex: A double-stranded DNA molecule or DNA-RNA hybrid, where each strand is of a different origin.

Homologous chromosomes: Chromosomes that have the same linear arrangement of genes—a pair of matching chromosomes in a diploid organism. See Chromosomes.

Homologous recombination: The exchange of DNA fragments between two DNA molecules or chromatids of paired chromosomes (during crossing over) at the site of identical nucleotide sequences.

Homozygote: An organism whose genotype is characterized by two identical alleles of a gene. See Allele, Genotype.

Insulin: A peptide hormone secreted by the islets of Langerhans of the pancreas that regulates the level of sugar in the blood.

Interferon: A family of small proteins that stimulate viral resistance in cells.

Intergenic regions: DNA sequences located between genes that comprise a large percentage of the human genome with no known function.

Introgression: Backcrossing of hybrids of two plant populations to introduce new genes into a wild population.

Intron: A noncoding DNA sequence within a gene that is initially transcribed into messenger RNA but is later snipped out.

Kilohertz (kHz): One thousand cycles per second.

Kilojoule: Standard unit of energy equal to 1000 joules, or 0.24 kilocalories.

Lag phase: The initial growth phase, during which cell number remains relatively constant prior to rapid growth. See growth phase.

Legume: A member of the pea family that possesses root nodules containing nitrogen-fixing bacteria.

Library, DNA: A large set of clones of DNA fragments from a particular organism. DNA libraries are usually maintained in *E. coli* or in bacteriophages.

Ligation: Joining two fragments of DNA end to end. See sticky ends.

Linked genes/markers: Genes and/or markers that are so closely associated on the chromosome that they are coinherited in 80% or more of cases.

Linker: A short, double-stranded oligonucleotide containing a restriction endonuclease recognition site, which is ligated to the ends of a DNA fragment.

Locus (plural = loci): A specific location or site on a chromosome.

Logarithmic phase (log or exponential growth phase): The steepest slope of the growth curve—the phase of vigorous growth during which cell number doubles every 20-30 minutes.

Lysogen: A bacterial cell whose chromosome contains integrated viral DNA.

Lytic: A phase of the virus life cycle during which the virus replicates within the host cell, releasing a new generation of viruses when the infected cell lyses.

Machine-phase chemistry: The chemistry of systems in which all potentially reactive moieties follow controlled trajectories (e.g., guided by molecular machines working in vacuum).

Macroscale: Larger than nanoscale; often implies a design that humans can directly interact with; too large to be built by a single assembler (one cubic micron of diamond contains 176 billion atoms).

Magnetostrictive material: A material that changes dimension in the presence of a magnetic field or generates a magnetic field when mechanically deformed.

Malignant: Having the properties of cancerous growth.

Manufacturing: The ability to make products, in this case ranging from clothing, to electronics, to medical devices, to books, to building materials, and much more.

Meat Machine: AKA Cabinet Beast. A box containing assemblers and raw material, within which is formed meat [or whatever else it was programmed to make].

Megabyte (MB): 220 (= 1,048,576, or about one million) bytes of information.

Megahertz (Mhz): One million cycles per second

Meiosis: The reduction division process by which haploid gametes and spores are formed, consisting of a single duplication of the genetic material followed by two mitotic divisions.

Metabolism: The biochemical processes that sustain a living cell or organism.

Metallothionein: A protective protein that binds heavy metals, such as cadmium and lead.

Micelles: Micelles are small, spherical structures composed of molecules that attract one another to reduce surface tension. The head of the molecule is hydrophilic, meaning it likes water, while the interior portion is hydrophobic, meaning it avoids water.

Microbial mats (biofilms): Layered groups or communities of microbial populations.

Microelectronics: Narrower terms: MEMS, nanoelectronics, optoelectronics, SED Single Electron Devices. Related terms: molecular electronics, semiconductors

Microencapsulation: Individually encapsulated small particles.

Microinjection: A means to introduce a solution of DNA, protein, or other soluble material into a cell using a fine microcapillary pipet.

Microsystem: A microscale machine that can sense information from the environment and act accordingly. Outside the U.S., it can also refer to microelectro-mechanical systems (MEMS).

Microtubule: A hollow cylindrical tube used for the transport of materials inside cells.

Miller indices: A set of 3 integers (4 for hexagonal) that designate crystallographic planes, as determined from reciprocals of fractional axial intercepts.

Millimeter: One thousandth of a meter, or about 1/26 of an inch.

Mitosis: The replication of a cell to form two daughter cells with identical sets of chromosomes.

Molecular - Nanotechnology: The technology of precisely-constructed molecular-scale machines; from nanometer: a billionth of a meter. For a good introduction, see What is Nanotechnology? or Ralph C. Merkle's nanotechnology page.

Molecular Beam Epitaxy: [MBE] Process used to make compound (multi-layer) semiconductors. Consists of depositing alternating layers of materials, layer by layer, one type after another (such as the semiconductors gallium arsenide and aluminum gallium arsenide).

Molecular biology: The study of the biochemical and molecular interactions within living cells.

Molecular genetics: The study of the flow and regulation of genetic information between DNA, RNA, and protein molecules.

Molecular markers: Genetic traits which are detectable on gels or blots and can be used to construct genetic maps. Molecular markers usually have no known function. RFLPs are molecular markers.

Molecular Wire: A molecular wire—the simplest electronic component—is a quasi-one-dimensional molecule that can transport charge carriers (electrons or holes) between its ends. [Michael D Ward]

Monoclonal antibodies: Immunoglobulin molecules of single- epitope specificity that are secreted by a clone of B cells.

Monoculture: The agricultural practice of cultivating crops consisting of genetically similar organisms.

Mycorrhizae: Fungi that form symbiotic relationships with roots of more developed plants.

Nanoscience: Nanoscience is the science of nanoscale phenomena. This is a very complicated area, since Quantum Mechanics is often needed to describe the physical properties of atoms and molecules properly.

Nitrocellulose: A membrane used to immobilize DNA, RNA, or protein, which can then be probed with a labeled sequence or antibody.

Nitrogen fixation: The conversion of atmospheric nitrogen to biologically usable nitrates.

Occupational Safety and Health Administration (OSHA): One of the U.S. agencies responsible for regulation of biotechnology. The major law under which the agency has regulatory powers is the Occupational Safety and Health Act.

Oncogenesis: The progression of cytological, genetic, and cellular changes that culminate in a malignant tumor.

Open pollination: Pollination by wind, insects, or other natural mechanisms.

Operator: A prokaryotic regulatory element that interacts with a repressor to control the transcription of adjacent structural genes.

Organelle: A cell structure that carries out a specialized function in the life of a cell.

Origin of replication: The nucleotide sequence at which DNA synthesis is initiated.

Paleontology: The study of the fossil record of past geological periods and of the phylogenetic relationships between ancient and contemporary plant and animal species.

Parasitism: The closee association of two or more dissimilar organisms where the association is harmful to at least one. See Commensalism, Parasitism, Symbiosis.

Particle bombardment: Using metal particles to blast DNA into cells for the purpose of genetic engineering. A gene gun is used to propel the metal particles.

Pheromone: A hormone-like substance that is secreted into the environment.

Phosphatase: An enzyme that hydrolyzes esters of phosphoric acid, removing a phosphate group.

Phosphodiester bond: A bond in which a phosphate group joins adjacent carbons through ester linkages. A condensation reaction between adjacent nucleotides results in a phosphodiester bond between 3' and 5' carbons in DNA and RNA.

Phosphorylation: The addition of a phosphate group to a compound.

Piezoelectric material: A ferroelectric material in which an electrical potential difference is created due to mechanical deformation, or conversely, in which the application of a voltage causes dimensional changes in the material.

Population: A local group of organisms belonging to the same species and capable of interbreeding.

Primary cell: A cell or cell line taken directly from a living organism, which is not immortalized.

Primer: A short DNA or RNA fragment annealed to single-stranded DNA, from which DNA polymerase extends a new DNA strand to produce a duplex molecule.

Prokaryote: A bacterial cell lacking a true nucleus; its DNA is usually in one long strand. See Eukaryote.

Pronucleus: Either of the two haploid gamete nuclei just prior to their fusion in the fertilized ovum.

Protease: An enzyme that cleaves peptide bonds that link amino acids in protein molecules.

Protein kinase: An enzyme that adds phosphate groups to a protein molecule at serine, threonine, or tyrosine residues.

Recessive gene: Characterized as having a phenotype expressed only when both copies of the gene are mutated or missing.

Recessive: Refers to the member of a pair of alleles that fails to be expressed in the phenotype of the organism when the dominant member is present. Also refers to the phenotype of an individual that has only the recessive allele.

Rhizobia: Bacteria in a symbiotic relationship with leguminous plants that results in nitrogen fixation. See Nitrogen fixation.

Rhizosphere: The soils region on and around plant roots.

Ribosome: Cellular organelle that is the site of protein synthesis during translation. See Organelle, Translation.

Ribosome-binding site: The region of an mRNA molecule that binds the ribosome to initiate translation.

Risk assessment: Defined as a formalized basis for the objective evaluation of risk in a manner in which assumptions and uncertainties are clearly considered and presented.

RNA: Ribonucleic acid. The three basic types are messenger RNA (mRNA), ribosomal RNA (rRNA), and transfer RNA (tRNA).

Roundup Ready: A trademark for plants that are genetically engineered to be resistant to the herbicide Roundup (technical name: glyphosate).

Salmonella: A genus of rod-shaped, gram-negative bacteria that are a common cause of food poisoning.

Satellite RNA (viroids): A small, self-splicing RNA molecule that accompanies several plant viruses, including tobacco ringspot virus.

Self-pollination: Pollen of one plant is transferred to the female part of the same plant or another plant with the same genetic makeup.

Semiconservative replication: During DNA duplication, each strand of a parent DNA molecule is a template for the synthesis of its new complementary strand. Thus, one half of a preexisting DNA molecule is conserved during each round of replication.

Sequence: Noun: the particular order of nucleotides in a DNA or RNA fragment. Verb: to determine the particular order of nucleotides in a strand of DNA.

Sequence-tagged site (STS): A unique (single-copy) DNA sequence used as a mapping landmark on a chromosome.

Sexual reproduction: The process where two cells (gametes) fuse to form one hybrid, fertilized cell. See Asexual reproduction, Gamete, Hybrid.

Signal transduction: The biochemical events that conduct the signal of a hormone or growth factor from the cell exterior, through the cell membrane, and into the cytoplasm. This involves a number of molecules, including receptors, pro- teins, and messengers.

Site-directed mutagenesis: The process of introducing spe- cific base-pair mutations into a gene.

Species: A classification of related organisms that can freely interbreed.

Spore: A form taken by certain microbes that enables them to exist in a dormant stage. It is an asexual reproductive cell. See Asexual reproduction, Dormant.

Structure-functionalism: The scientific tradition that stresses the relationship between a physical structure and its function, for example, the related disciplines of anatomy and physiology.

Supergene: A group of neighboring genes on a chromosome that tend to be inherited together and sometimes are functionally related.

Supernatant: The soluble liquid &action of a sample after centrifugation or precipitation of insoluble solids.

Susceptibility: When individual or species will become ill or possibly die in the face of the stressful event or circumstances.

Upstream: The region extending in a 5' direction from a gene.

Vaccinia: The cowpox virus used to vaccinate against smallpox and, experimentally, as a carrier of genes for antigenic determinants cloned from other disease organisms.

Variable surface glycoprotein (VSG): One of a battery of antigenic determinants expressed by a microorganism to elude immune detection.

Variation: Differences in the frequency of genes and traits among individual organisms within a population.

Vasculocyte: In medical nanorobotics, a theorized (nanorobotic) device capable of performing repairs of an injured vascular luminal surface.

Vasculography: A physical description (and mapping) of the human vascular system.

VC (Venture capitalist): A person or organisation that invests in promising companies during early development stages.

Vector: In cell biology, an agent (virus or plasmid) used to transmit genetic material to a cell or organism.

1

Biotechnology Research: A National Achievement Survey

Biotechnology in Australia

- The Australian Bureau of Statistics (ABS) added 10 questions on biotechnology to its Businesses Survey of Research and Experimental Development for reference year 2003-04. The survey excludes firms mainly engaged in Agriculture, forestry and fishing. The survey included the OECD definition of biotechnology.
- The Businesses Survey of Research and Experimental Development is an annual mandatory survey. Non-responding firms which reported R&D activity in the previous cycle have data imputed based on the previous response. The response rate for the 2003-04 survey was 86%.
- In 2003-04, 304 firms were directly or indirectly involved in biotechnology R&D.

Two hundred and twenty-seven firms performed intramural biotechnology R&D, while 78 of these firms also paid another organisation to conduct biotechnology R&D on their behalf. Another 77 firms only contracted out biotechnology R&D to another organisation (firm, university, etc).

- The intramural biotechnology R&D by the 227 firms amounted to PPP$ 200.5 million or 3.8% of total BERD.
- For greater comparability with the other country profiles, the results for R&D applications and sectors focus on the intramural biotechnology R&D.
- When firms were asked to classify their intramural R&D to the appropriate biotechnology application, 47% classified themselves as active in the Human Health bioindustry sector, 23% in the Ag-biotech sector, and 13% in Industrial processing.
- In terms of biotechnology R&D expenditure, 69% was allocated to the Human Health bioindustry sector, followed by 12% in Agbiotech.
- The sector Property and Business Services (ANZSIC Division L) reported the largest share of biotechnology-active firms (61%) and the highest share of biotechnology R&D expenditures (69%).
- One hundred and forty-five firms paid one or more Australian organisations to perform biotechnology-related R&D. Another 36 contracted out biotechnology R&D abroad.
- Seventy-three percent of the 145 firms that contracted biotechnology R&D to Australian organisations had fewer than 19 employees.
- Smaller firms, with 99 or fewer employees, favoured contracting biotechnology R&D to Universities or other higher education institutions.
- Firms with over 100 employees favoured contracting biotechnology R&D to the Commonwealth Scientific and Industrial Research Organisation (CSIRO) and Contract Research Organisations (CROs).

Biotechnology in Belgium

- A Belgian Biotechnology Use and Development survey was run for the reference year 2003 at the initiative of the Office for Scientific, Technical and Cultural Affairs. The practical organisation of the survey " data collection and analyses " was undertaken with several partners: the *Limburgs Universitair Centrum* (Center for Statistics, ITEO), the University of Liege (CRGB) and the Vlerick Management school.
- This survey was the first of its kind in Belgium. The survey was modeled on the recommendations of the OECD *Ad hoc* Biotechnology Statistics group and included the OECD definition of biotechnology.
- The survey was voluntary. The survey response rate was 31% and there was no weighting for non-respondents.
- The survey identified 73 biotechnology firms in Belgium in 2003.
- The European Classification of Economic activities (NACE) was used to categorise firms into different sectors.
- In 2003, firms were classified into four industrial sectors: Agrofood (a combination of Agriculture, and Food and Beverages), Pharmaceuticals, R&D and Business Services and Human Health Services. There was also an "Other" category.
- Twenty-four firms (33%) fell in the "Other" category; 17 firms (23%) were classified in the Pharmaceuticals sector; 14 firms (19%) were classified in R&D and Business Services and 7 firms (10%) were in Human Health Services.
- In 2003, 41% of biotechnology firms had less than

10 employees. Only 7% had over 500 employees.

- Seventy of the 73 biotechnology firms responded to the detailed employment questions in the survey.
- In 2003, the 70 biotechnology firms employed 11,137 persons. Of these, 4,261 (38%) had biotechnology-related responsibilities, including 1,984 with biotechnology R&D responsibilities (18% of the total employment).
- A much higher percentage of the total employment in biopharma versus nonbiopharma firms had biotechnology responsibilities: 51.5% compared to 19.5%.
- The 2003 survey also collected information on the nine types of biotechnology techniques used by firms, and whether firms were using, or planning to use, an existing technique or were developing a new technique.
- The top biotechnology techniques developed by firms were Proteins and molecules and Process technologies, with 36.1% of all firms reporting developing each of these techniques.

Cell tissue culture techniques ranked second with 34.7% of all firms reporting developing that technique.

- The top three biotechnology techniques used by firms were: Proteins and molecules (used by 54.2% of all firms), DNA coding (used by 47.2% of all firms) and Process technologies (used by 44.4% of all firms).
- Nanotechnologies and the 'other' group were the least used and developed biotechnology techniques by firms.

Biotechnology in Canada

- Statistics Canada has been collecting data on biotechnology since 1989. For a detailed account of the past surveys, refer to the "OECD Biotechnology Statistics Inventory".
- Canadian biotechnology data is derived from two sources: the Federal Science Expenditures and Personnel Survey and the Biotechnology Firm Survey (1997), which was replaced by the Biotechnology Use and Development Survey (1999, 2001, 2003).

Federal Science Expenditures and Personnel Survey

- The Federal Science Expenditures and Personnel Survey has been collecting data on federal biotechnology-related activities annually since fiscal year 1996-97. These data relate to science and technology (S&T) expenditures, which include R&D and related scientific activities (RSA), and to full-time equivalent (FTE) personnel associated with biotechnology activities.
- In 2003-04, the federal government spent PPP$ 575.9 million on biotechnology R&D, which represented 13% of total federal R&D expenditures. Related scientific activities accounted for an additional PPP$ 23.3 million in federal government expenditures in 2003-04.
- The higher education sector was the largest recipient of biotechnology funds with PPP$ 304.5 million (52%) of all funds. Thirtytwo percent of biotechnology funds were intramural, *i.e.* spent by federal government research institutes and laboratories in Canada.
- From 1997 to 2003-04, federal government biotechnology expenditures grew by an average of 18% per year.

Biotechnology Firm Survey and Biotechnology Use and Development Survey

- Statistics Canada has run four surveys of biotechnology firms: for 1997, 1999, 2001 and 2003. All surveys provided a definition of biotechnology. As of 2001, the OECD definition of biotechnology was provided.
- All firm surveys were conducted on a voluntary basis and weighting was used to reflect the entire firm population. In 2003, the survey response rate was 80%.
- The surveys' target population was "innovative biotechnology firms," defined as firms that meet at least one of three criteria: "it has one or more biotechnology products or processes on the market, it is currently developing products or processes that require the use of biotechnology, or it considers biotechnology central to its activities or strategies." The 2003 survey excluded firms with less than 5 employees, less than PPP$ 80,321 in R&D expenditures, and with less than PPP$ 200,803 in sales. The survey also excluded not-forprofit organisations, universities, government laboratories, hospitals, companies that use only traditional biotechnologies, and service sector firms.
- http://bioportal.gc.ca/trends.
- In 2003, there were 490 innovative biotechnology firms in Canada, compared to 282 in 1997. The number of biotechnology firms grew by 74% from 1997 to 2003.
- In all the years surveyed, the majority of biotechnology firms were Small and Medium- Sized Enterprises (SMEs), which are defined as firms

with between 5 and 149 employees by Statistics Canada. In 2003, 88% or 429 of all biotechnology firms were SMEs.

- In 2003, 262 biotechnology firms (54%) were classified in the Human health sector. In 1997, 48% of biotechnology firms were classified in Human health.
- In 2003, biotechnology firms spent PPP$ 1,194.4 million on biotechnology R&D, or 65% of their total R&D spending. Spending on biotechnology R&D grew by 192% from 1997 to 2003. Total R&D spending by the biotechnology firms grew at a slower rate of 139% over the same period.
- Biotechnology revenues were PPP$ 3,085.9 million or 13% of the firms' total revenues in 2003. In 1997, biotechnology revenues constituted only 6% of the total revenues of biotech firms.
- In 2003, biotechnology firms employed 75,448 employees, of which 11,863 (16%) had biotechnology-related responsibilities. A little over half, or 6,441, of the employees with biotechnology-related responsibilities were in scientific research: "Scientific research and direction" and "Technicians".
- In 2003, biotechnology firms reported that they had 17,065 biotechnology products/processes at various stages of development including: R&D, Pre-clinical trials/confined field trials, Regulatory phase/ unconfined release and Approved/On market/In production.
- In 2003, biotechnology firms reported that 65% of biotechnology products/processes were on the market, 11,046 of 17,065. In 1999, only 38% of biotechnology products/processes were on the market, while 49% were in the R&D stage.

- In 2003, the Human health sector had the largest number of biotechnology products/processes, with 10,692 or 63%. The Agriculture biotechnology sector had the second largest number of biotechnology products/processes in 2003, with 4,813 or 28%, and Food processing was third, with 622 or 4%.
- The 2003 results show several changes from 1999, where the Bioinformatics sector had the largest number of biotechnology products/processes, with 7,249 or 41%. The Agriculture biotechnology sector had the second largest number of biotechnology products/processes in 1999, with 5,557 or 32%, and Health was third, with 3,435 or 20%.

Biotechnology in Shanghai, China

- In 2004, the Shanghai Science and Technology Commission undertook a biotechnology survey of Shanghai, for the Ministry of Science and Technology of the People's Republic of China.
- Shanghai had an estimated population of 17.1 million inhabitants in 2003. For comparison, Belgium had an estimated 10.5 million6 inhabitants in 2005.
- This mandatory survey, covering reference year 2003, was the first of its kind.
- The survey provided both a single definition and a list-based definition of biotechnology.
- The survey scope covered firms, R&D institutions, and higher education and subsidiary institutions.
- The survey focused on 'modern' biotechnology and did not include traditional biology-related companies.
- In 2003, 158 firms, 31 R&D institutions and 22 higher education and subsidiary institutions were active in biotechnology in Shanghai.

- Thirty-three percent of biotechnology firms were in the R&D stage of activity and 20% were involved in product & process development.
- Over three-quarters of all biotechnology firms were in the manufacturing sector (123 firms).
- Three-quarters of all biotechnology firms had less than 150 employees (119 firms).
- The dominant sector of application was biomedicine (66 firms), followed by human health (34 firms).
- In 2003, intramural biotechnology R&D by firms was estimated to be PPP$ 204.5 million.

Over three-quarters of this R&D was spent in the manufacturing sector.

- In 2003, firms reported having 1,447FTE employees with biotechnology R&D-related duties.
- Biotechnology firms reported having 388 products in the pipeline or on the market. Over half of these products were pharmaceuticals (206 of 388).

Biotechnology in Denmark

- The Danish Centre for Studies in Research and Research Policy added a biotechnology question to its private sector and public sector R&D surveys in 1991. In both surveys, respondents were asked to estimate the percent of total R&D expenditures devoted to biotechnology. Neither survey included a definition of biotechnology.
- The private sector R&D survey covers the private business enterprise sector and the public R&D survey covers institutes, universities, hospitals and non-profit organisations.
- The private sector R&D survey is voluntary and is

conducted every other year. The public sector R&D survey is annual and mandatory.

The response rate for the 2003 private survey was 63% and weighting was conducted for non-respondents. The response rate for the public survey was 95% and no weighting was conducted for non-respondents.

- The Danish Centre for Studies in Research and Research Policy took the biotechnology R&D survey results and created a database of biotechnology firms in Denmark. The database was supplemented with firms identified using other sources, including biotechnology organisations, science parks etc.
- Firms with biotechnology activities were then categorised into two main groups: core biotechnology firms and firms with significant biotechnology activities. Both groups are active in biotechnology R&D.
- Core biotechnology firms are those whose primary activity is biotechnology. Firms with significant biotechnology activities are those where biotechnology is not their primary activity.
- The creation of the database was only possible using data beginning in 1997; earlier data are not comparable.
- For more information on the methodology undertaken by the Danish Centre for Studies in Research and Research Policy a study was released in April 2004 and is available on line, at: http://www.cfa.au.dk/Publikationer/Working_p apers/WP2004_1.pdf.
- In 2003, there were 181 core biotechnology firms and 86 firms with significant biotechnology activities in Denmark.

- From 1997 to 2003, the number of core biotechnology firms grew at an average annual rate of approximately 20%. Over the same period, the number of firms with significant biotechnology activities grew at a slower rate (4.5%).
- Most core biotechnology firms were started in, or after 2000.
- In 2003, the private sector spent PPP$ 726.8 million on biotechnology R&D, or 23.8% of total private R&D spending. The public sector spent PPP$ 131.3 million on biotechnology R&D, which constituted 9.9% of total public R&D spending.
- From 1997 to 2003, extramural biotechnology R&D spending (22.6% average annual growth rate) grew faster than intramural spending (12.4% average annual growth rate).
- In 2003, private sector biotechnology firms employed 4,781 full-time equivalent (FTE) R&D personnel (16.5% of total private sector R&D FTEs). In the same year, the public sector employed 1,406 FTE biotechnology R&D personnel (9.5% of total public sector R&D FTEs).
- From 1997 to 2003, the average annual growth rate of private biotechnology R&D employment was 5.2%; it was 0.4% for the public sector.

Biotechnology in Finland

- In Finland, two agencies have collected data on biotechnology: Statistics Finland and ETLA, the Research Institute of the Finnish Economy.

Statistics Finland

- Statistics Finland added a question on biotechnology R&D to their 2003 R&D survey. The survey included the OECD definition of biotechnology.

- The survey response rate was 83% and the results are weighted for non-respondents.
- In 2003, total R&D expenditure on biotechnology by both the public and business sectors was PPP$ 192.9 million, or 3.7% of all R&D expenditure (PPP$ 5,220.3 million). The business sector spent the most on biotechnology R&D (PPP$ 88.3 million). However, this represented only 2.4% of total business enterprise R&D expenditures. The higher education sector spent 8.6% or PPP$ 87.4 million on biotechnology R&D, out of a total higher education sector R&D expenditure of PPP$ 1,015.4 million. The government sector spent PPP$ 17.3 million on biotechnology R&D, or 3.2% of total government R&D.
- In 2003, the 102 firms engaging in biotechnology R&D employed 2,394 persons, 13% of whom had PhDs. The proportion of personnel with PhDs is higher for biotechnology firms than for other R&D active firms, where only 5% had PhDs.
- In 2003, 44 of the biotechnology R&D active firms were in the Manufacturing sector and 34 in the Research & Development sector (Services sector ISIC 73).
- Sixty-six percent of all biotechnology R&D active firms had 49 or fewer employees.

Research Institute of the Finnish Economy

- ETLA conducted two surveys on the Finnish biotechnology industry, one in 2002, covering 2001, and another in 2004, covering 2003. At the end of 2001, there were an estimated 119 biotechnology firms in Finland, including 73 Small and Medium-sized Enterprises (SMEs). At the end of 2003, there were approximately 120 biotechnology firms in

Finland, of which 93% were SMEs (112 firms). The survey included the OECD definition of biotechnology. The survey response rate was 71% and there was no weighting for nonrespondents.

- The 2004 survey focused only on SMEs, as it was felt that the larger firms would skew results.
- In 2003, the Diagnostics and devices sector had the largest number of firms (43 SMEs), followed by the Drug discovery/development sector (34 SMEs), and then the Food and feed sector (22 SMEs). Some SMEs were active in more than one sector and are therefore doublecounted.
- Total sales by Finnish biotechnology SMEs were approximately PPP$ 345.3 million in 2003. However, the sector as a whole was not making a profit, with total operating profits and total net profits at PPP$ -62.4 million and PPP$ -72.8 million, respectively.
- Sales in Enzymes were highest, representing 34% of total sales by Finnish biotechnology SMEs in 2003. The Drug development sector followed with 24% in sales. The Food and feed sector ranked third, with 20% of sales. The Bioinformatics sector had the lowest sales figures, with 0.2% of sales.
- The 2004 survey also collected information on the cross between the biotechnology technique (using the OECD list-based definition) and sector of application.
- DNA biotechnology techniques were most prevalent. These techniques were used in every sector of application.
- Only three techniques " all process biotechnology techniques "were not used at all by Finnish biotechnology SMEs: biopulping, biodesulphurisation and bioremediation.

- The bioinformatics sector reported using the widest array of techniques, followed closely by the drug development sector and the diagnostics sector.
- See the ETLA study released in April 2005 for more information. It is available on line at: http://www.etla.fi/files/1302_Dp978.pdf.

Biotechnology in France

- The MENRT, the Ministry of Education, Research and Technology (*Ministère de l'éducation nationale, de la recherche et de la technologie*), first added a biotechnology question to its annual business enterprise R&D survey in 2000. Respondents were asked to estimate the percent of total R&D expenditures devoted to biotechnology. The survey included the OECD definition of biotechnology.
- The French R&D survey is mandatory and is corrected for non-response. The response rate to the 2003 R&D survey was 71.5%.
- In 2003, 755 firms undertook biotechnology R&D in France, which represented 11% of all firms undertaking R&D.
- Fifty-seven percent of all R&D active pharmaceutical firms were undertaking biotechnology R&D. Forty-three percent of all R&D active agriculture firms were undertaking biotechnology R&D.
- In 2003, the 755 firms spent PPP$ 1,342.0 million on biotechnology R&D, which represented 5.6% of total business enterprise R&D spending.
- Biotechnology R&D was concentrated in 458 firms that dedicated over 75% of their total R&D expenditure to biotechnology. These firms are defined here as 'core' biotechnology firms.

- The core biotechnology firms spent PPP$ 1,196.6 million on biotechnology R&D, which represented 89% of total biotechnology R&D.
- In 2003, 95% (433) of the core biotechnology firms were SMEs with less than 250 employees.
- There were 170 core biotechnology firms (37% of all core biotechnology firms) where over 75% of the employees were working on R&D.

These firms spent PPP$ 704.7 million (59% of total R&D spending) on biotechnology R&D in 2003.

- In 2003, most core biotechnology firms were in the Pharmaceuticals sector (255 of 458). These firms were responsible for 80% (PPP$ 958.5 million) of all biotechnology R&D.
- Over half of all R&D-active pharmaceutical firms (57%) in France undertook biotechnology R&D in 2003. These firms spent PPP$ 1,010.3 million on biotechnology R&D, which represented 75% of total biotechnology R&D spending.

Biotechnology in Germany

- The German Federal Statistical Office has run three Biotechnology firm surveys, for reference years 2000, 2002 and 2004. All three surveys used the OECD definition of biotechnology.

The results presented here are limited to the 2002 and 2004 surveys and were weighted to reflect the total population of biotechnology firms.

- The surveys were conducted on a voluntary basis and the overall response rates were 58% in 2002 and 65% in 2004.
- The survey target population was divided into five categories7: " Category I: Core biotech firms which

mainly apply modern biotechnological procedures (538 firms); " Category II: Suppliers to the core biotech firms and research institutions (375 firms); " Category III: Firms that are both core biotech firms and suppliers (38 firms); " Category IV: Large life sciences firms, or their subsidiaries, with a significant focus on biotechnological activities (31 firms); and " Category V: Consultants and financial service providers to biotech (177 firms).

- Categories I, III, and IV combined cover biotech activities, including R&D, that are included in the OECD definition. No further details are given for categories II and V because they are rarely included as biotech firms in other countries, which usually report activities for 'core' biotech firms only or for all firms active in biotech R&D.
- All employment, R&D, and turnover in category I is assumed to be relevant to biotech.

Conversely, data for category IV are limited to biotech activities: the R&D data exclude nonbiotech R&D. For category III, all employment and R&D is assumed to be relevant to biotech, but the results for turnover exclude non-biotech sales. The results exclude 49 firms that were not in operation at the time of the survey, including: 34 firms in category I (core), 7 firms in category II (suppliers), 2 firms in category IV (large life sciences) and 6 firms in category V (services).

- In 2004, there were 607 firms involved in biotech categories I, III and IV combined.

Eighty-nine percent of these firms were core biotech firms (538 firms).

- Most core biotech firms in 2004 had less than 100 employees (519 firms or 96%).

- In 2004, there were 24,134 biotech-active employees, of which 11,958 (50%) were employed by core biotech firms and 10,995 (46%) by large life science firms.
- There were an estimated 8,024 biotech R&D employees in 2004, of which 5,438 (68%) were in core biotech firms. Smaller firms had a higher share of employees active in R&D.
- From 2002 to 2004, total biotech employment grew only in the large life sciences firms (16%). Employment fell in the core biotech firms (- 10%) and in the core/supplier firms (- 55%).
- In 2004, an estimated PPP$ 1,346.8 million was spent on biotech R&D: PPP$ 791.7 million (59%) by the core biotech firms and PPP$ 533.7 million (40%) by the large life science firms.
- Total turnover, from biotech sales only, was PPP$ 3,837.5 million, of which PPP$ 2,622.5 million (68%) was by large life science firms and PPP$ 1,107.0 million (29%) by core biotech firms.
- From 2002 to 2004, biotech R&D fell by 30% for core firms and by 10% for large life sciences firms. During the same period, biotech turnover increased by 11% for core firms and 5% for biotech products in large life science firms.
- The firms were asked to classify their activities into three fields: "green" biotech, which covers agricultural and food biotech; "red" biotech, which covers health (human and animal) and bioinformatics; and "grey" biotech, which covers industrial and environmental biotech.
- Most core biotech firms classified their activities as "red" biotech (67%).

Biotechnology in Iceland

- In 2003, the Ministry of Education, Science and Culture commissioned RANNIS, The Icelandic Centre for Research, to run a one-off survey to measure R&D in biotechnology.
- The survey covered 2001 to 2003 and was the first of its kind in Iceland. The survey achieved an overall response rate of 100%.
- The survey provided both the OECD single and list-based definitions of biotechnology.
- The survey covered: the business enterprise sector, the government sector, the higher education sector and the private non-profit sector.
- Combined, the sectors spent PPP$ 73.1 million on biotechnology R&D in 2003. Ninety-two percent of this amount was undertaken by the business enterprise sector.
- In 2003, 23 firms were undertaking biotechnology R&D. These firms spent PPP$ 67.2 million on biotechnology R&D.
- In 2003, total business enterprise R&D expenditure by all firms in Iceland was PPP$ 130.7 million, over half of which was spent on biotechnology R&D.
- Source of funding data shows that 86% of business enterprise R&D funds were financed from abroad in 2003.
- Combined, the biotechnology-active sectors employed 1,484 employees (headcount data). This figure represents total employment by the sectors, as no data was collected for biotechnology employment alone. Over half of these employees were in the business enterprise sector (969 or 65%).

- Six hundred and forty-two employees had R&D-related duties (headcount data). The business enterprise sector had the largest share of R&D biotechnology personnel (71%).

Combined, the government sector, the higher education sector and the private non-profit sector had 29% of all R&D personnel.

- When calculated in full-time equivalents, 568.1 employees worked on biotechnology R&D in 2003. Seventy-four percent of these biotechnology R&D FTEs were in the business enterprise sector (422 FTEs).
- Over half of all biotechnology R&D FTEs were women (58% or 330.2 women).
- In 2003, Diagnostics and Therapeutics were the main activities for the largest number of performers. Silviculture is the only activity with no performers.
- Health-related activities were the main focus of biotechnology R&D in 2003, with 88% of total biotechnology R&D dedicated to health (PPP$ 64.1 million). Ninety-seven percent of this amount came from the business enterprise sector.

Biotechnology in Israel

- In 2004, the Israeli Central Bureau of Statistics ran a survey on biotechnology firms in Israel for 2002. The survey used the OECD definition of biotechnology.
- The survey was mandatory and had a response rate of 95.5% and the results were weighted for non-respondents.
- In 2002, 148 firms were engaged in biotechnology in Israel.

- In 2002, 117 (79%) of all biotechnology firms were in the Research & Development sector (Services sector ISIC 73), which included research institutes and start-up companies which did not yet manufacture biotechnological products, but were in various stages of product development.
- Most of the biotechnology firms in Israel were small, 76% had 20 or fewer employees in 2002.
- In 2002, almost half of the firms were engaged in Human health care applications, with 73 firms or 49%. The second largest application was Agriculture and marine biotechnology, with 28 firms or 19%. The Environment and afforestation application was third, with 24 firms or 16%.
- In 2002, expenditure on biotechnology R&D was an estimated PPP$ 251.1 million. This represented 4.9% of total Business Enterprise R&D expenditure (PPP$ 5,095.3 million).
- Firms active in Human health applications were responsible for 61% (PPP$ 152.1 million) of all biotechnology R&D in 2002. Bioinformatics had the second largest share of total biotechnology R&D expenditures, with 15% or PPP$ 37.6 million.
- Biotechnology firms employed 3,892 persons, 3,427 (88%) of whom worked on biotechnology-related tasks. A quarter of these employees had PhDs (25%), while 39% had Bachelor's and Master's degrees. Of the employees working on biotechnology-related tasks 47% were in R&D.
- In 2002, 55% of all biotechnology employees worked in Human health applications (1,879 employees).
- In 2002, biotechnology firms generated PPP$ 331.8 million in sales. Fifty-two percent of biotechnology

sales came from Human health applications (PPP$ 174.0 million).

- Seventy-five percent of biotechnology sales is generated from exports (PPP$ 250.5 million).
- The biotechnology firms raised PPP$ 186.1 million from external sources of finance, of which 96% was raised by the R&D sector.
- Venture capital accounted for 5.6% of the total funding from external sources.

Biotechnology in Italy

- The National Institute of Statistics (ISTAT) added a question on biotechnology R&D to their annual business enterprise R&D survey as of 1991. As of 2002, the survey included the OECD definition of biotechnology.
- The business enterprise R&D survey is a mandatory census-based survey. The survey does not apply any size cut-off for R&D performers.
- In 2003, the overall rate of response was 49.6%. The results of the Italian business R&D survey are not weighted for non-response.
- In 2003, 172 firms reported performing biotechnology R&D in Italy.
- The European Classification of Economic activities (NACE) was used to categorise firms into different sectors.
- In 2003, 67 of the biotechnology R&D active firms were in the Manufacturing sector and 62 in the Services sector.
- In 2003, the largest number of biotechnology R&D active firms was in the Research & Development

sector, with 42 firms, followed by the Chemical industry, with 36 firms.

- In 2003, total R&D expenditure on biotechnology amounted to PPP$ 236.2 million, or 2.8% of all business enterprise R&D expenditure (PPP$ 8,309.8 million).
- Forty-four percent of all biotechnology R&D expenditure was undertaken by firms in the Research & Development sector. Firms in the Chemical Industry sector, which includes Pharmaceuticals, undertook 20% of all biotechnology R&D.
- The Health related services sector had the largest proportion of biotechnology R&D over total R&D (39.4%), however, this represented only 1.4% of total biotechnology R&D expenditure.
- R&D on process biotechnology techniques (bio-reactors, fermentation, bio-catalysis, industrial bio-processes, and environmental bio-technologies) accounted for 38% of biotechnology R&D expenditures. All but the sector health-related services conducted R&D on process biotechnology.
- The Research & Development sector reported using the widest array of biotechnology techniques.

Biotechnology in Japan

- The Japanese Bioindustry Association (JBA) has been running a Biotechnology Industry Survey since 1997. Since 2000, the survey has been conducted under contract by the Ministry of Economy, Trade and Industry (METI). This has increased the survey response rate significantly: in 1999 the response rate was 37.9%, and in 2003 it was 75.7% with responses received from 1,162 firms; there was no weighting for non-respondents.

- The survey scope covers "traditional and/or modern biotechnology". The former includes traditional fermentation, cultivation, mutagenesis and pollution treatment technology which explains why the data are skewed towards the Food/drink sector. Modern biotechnology includes recombinant DNA, cell fusion, tissue culture and biomimetic technology.
- The largest share of Japanese firms were in the Food or drink manufacture sector (21%), followed by the Chemical industry sector (11%) and the Pharmaceuticals sector (11%).
- Sixty-one percent of the firms had less than 300 employees.
- Conventional Fermentation, Cultivation and Mutagenesis Technologies accounted for 77% of total production.
- Traditional biotechnology products accounted for 82% of biotechnology production, while modern biotechnology products accounted for 18%. Production is measured as the value of domestic shipments, which was PPP$ 45,626.3 million for traditional biotechnology products and PPP$ 9,886.3 million for modern biotechnology products.
- Food products accounted for 62% of total production by biotechnology firms and for 75% of traditional biotechnology shipments.
- Pharmaceuticals, Diagnostic Reagents & Medical Instruments represented 58% of all products using modern biotechnology.
- Domestic production of modern biotechnology products increased from PPP$ 8,359.2 million in 2000 to PPP$ 9,886.3 million in 2003. The share of total modern biotechnology production from rDNA-

based products increased from 36.2% in 2000 to 39.9% in 2003.

Biotechnology in Korea

Biotechnology Survey

- The Korean Ministry of Commerce, Industry and Energy (MOCIE) has been running an annual biotechnology firm survey since 2002.
- This survey is conducted on a voluntary basis and achieved an overall response rate of 100%.
- This survey, which is limited to the manufacturing sector, provides a single definition of biotechnology and uses the Korean Bio-industry classification system.
- The Korean Bio-industry classification system classifies biotechnology firms into eight sectors: Biopharmaceutical, Biochemical, Biofood, Bioenvironmental, Bioelectronics, Bioprocess and equipment, Bioenergy and bioresource, and Bioassay, bioinformatics and R&D services.
- The survey focused on 'modern' biotechnology.
- In 2004, there were 640 firms active in biotechnology in Korea.
- Thirty percent of these firms were in the Biopharmaceutical sector, followed by 25% in the Biofood sector.
- In 2004, 69% or 422 of all biotechnology firms were small, with less than 50 employees.
- Biofoods dominated production by biotechnology firms (43%), followed by Biopharmaceuticals (40%).
- In 2004, biotechnology firms reported employing 12,138 employees. This figure represents total

employment by the 640 firms, as no data was collected for biotechnology employment alone.

- The largest share of employees worked in the Biopharmaceutical sector (36%), followed by the Biofood sector (29%).
- Over half of the employees had R&D-related duties (54% or 6,554 employees).
- The firms were asked to classify the biotechnology they used within 13 categories: Genetic engineering, Protein engineering, Other macromolecule engineering, Cell and tissue engineering, Systems biology and bioinformatics, Metabolic engineering, Bioprocess, Bioresource production and utilisation, Environmental biotechnology and bioenergy technology, Nanobiotechnology, Bioelectronics, Biosafety and bioefficiency, and Other biotechnology.
- In 2004, Bioprocess technology was used most frequently. Twenty-four percent of all biotechnology firms used it in the R&D stage, and 14% in the production stage.

Research and Development Survey

- The data that follow are not based on the same survey; the firms referred to below are not the same 640 biotechnology-active firms referred to above.
- Combined, firms, public research institutes and universities spent PPP$ 1,663 million on biotechnology R&D in 2004. These data are based on the results of the 2004 Survey of R&D in Science and Technology run by the Korean Ministry of Science and Technology.
- Firms undertook the largest share of R&D expenditure (42%), followed by universities (36%) and public research institutes (22%).

- According to the Korean Ministry of Science and Technology, the Korean government invested PPP$ 1,186.6 million on biotechnology R&D in 2005.
- The average annual growth rate of government biotechnology R&D expenditures was 27% from 2001 to 2005.

Biotechnology in New Zealand

Biotechnology Survey

- Statistics New Zealand has run biotechnology surveys in 1999, 2004 and 2005. Only the last two surveys adopted the OECD definition of biotechnology. Therefore, no data from the 1999 survey are given because they are not comparable with the 2004 and 2005 results.
- The biotechnology survey is mandatory and covers both the public and private sectors. The overall response rates were 94% for 2004 and 93% for 2005. No imputation was conducted.
- The surveys' target population, referred to as organisations, includes: firms, institutions receiving public funding for biotechnology R&D, Crown Research institutes, universities, microbiology units in metropolitan hospitals and the New Zealand Blood Service.
- Full details on the 2004 survey methodology and results are available on line at: http://www.stats.govt.nz/NR/rdonlyres/FDD38 C83-2F57-42C1-AE2D- 84CCD7D4D143/0/BiotechNZ2004.pdf. Full results for the 2005 survey are forthcoming.
- In 2004, 129 biotechnology organisations had used one or more biotechnology techniques in the last three years.

- In the 2005 survey the period under review was reduced to two years. In 2005, 135 biotechnology organisations had used one or more biotechnology techniques in the last two years while 126 organisations were currently using or intended to use one or more biotechnology techniques in the next two years.
- In 2005, 84 of 126 organisations were in the private sector, the remaining 42 were in the public sector.
- In 2005, 50% of firms reported using one or more biotechnologies for applications in 'Innovative foods & human nutrition' and 46% for 'Biomedical science and drug discovery'.
- In the public sector, 71% of organisations used one or more biotechnologies for applications both in 'Impacts & integration of emergent technology' and for 'Environmental technologies'.
- Most biotechnology techniques were used in the R&D stage, 62% in 2004 and 66% in 2005.
- Of the total 750 biotechnology techniques used in all stages (R&D, part of production process, part of product sold), 52% are due to the public sector in 2004.
- In 2005, the biotechnology sector employed 2,424 people, of which 34% had PhDs and 39% had postgraduate or undergraduate degrees (headcount data). The private sector employed 38% of all employees (918 employees) in 2005.
- Although total biotechnology employment grew by 7% from 2004 to 2005, there was an 8% fall in the number of employees with PhDs.

Technical and trade employees grew by 92% from 183 in 2004 to 352 in 2005.

- In 2004, PPP$ 288.5 million was spent on biotechnology. Biotechnology organisations generated PPP$ 452.8 million in income.
- In 2005, expenditure on biotechnology increased to an estimated PPP$ 344.3 million but income fell slightly to PPP$ 449.5 million.

Research and Development Survey

- Statistics New Zealand included a question on biotechnology in their 2004 R&D survey covering business, government and university spending on R&D. This survey included the OECD definition of biotechnology.
- The 2004 R&D survey is mandatory and the overall response rate was 84%. Results were weighted for non-respondents.
- Biotechnology R&D was performed in 2004 by 116 businesses, 11 government sector institutes, and 8 universities.
- In 2004, almost 23% of total R&D was allocated to biotechnology. Just over 20%, PPP$ 94.9 million, of all business enterprise R&D was for biotechnology R&D. The higher education sector allocated PPP$ 57.5 million (19%) to biotechnology R&D and the government sector PPP$ 91.2 million (30%) of its R&D budget to biotechnology.

Biotechnology in Norway

Higher Education and Research Institute sectors Research and Development Survey

- The Norwegian Institute for Studies in Research and Higher Education (NIFU) collects R&D data for the higher education sector and the research institute sector – both government institutes and institutes serving the business enterprise sector. In

2001 and 2003 NIFU included a special questionnaire on biotechnology R&D – using the OECD definition of biotechnology – within their regular R&D survey.

- The R&D survey for the higher education sector and the research institute sector is a biennial survey run on a voluntary basis. The response rate to the 2003 survey was 80%. In general, results were corrected for nonresponse, however the results from the biotechnology R&D questions were not corrected.
- This special questionnaire covered 82 institutes in the higher education and 22 institutes in the research sector. Fourteen of the 22 research institutes were classified in the government sector, the remaining eight were research institutes serving enterprises.

Business Enterprise Research and Development Survey

- Statistics Norway, the agency that collects R&D data for the business enterprise sector, added a question on biotechnology to their R&D survey in 1985. The OECD definition was used as of the 2003 R&D survey.
- The business enterprise R&D (BERD) survey is run on an annual basis. The biotechnology question however, is only included biennially.
- The survey is mandatory. The response rate to the 2003 BERD survey was 95%. For small enterprises, results are weighted for nonrespondents.
- For the very small number of nonrespondents among larger enterprises, other information is used to create estimates. The survey excludes firms with less than 10 employees.
- In 2003, biotechnology R&D expenditure in the Norwegian higher education, institute and business enterprise sectors amounted to PPP$ 119.5 million,

or 4% of total R&D expenditure (PPP$ 2,964.4 million). About 56% of these R&D funds were spent by the higher education sector, 25% by the industrial sector, and the remainder by the institute sector (20%).

- In 2003, R&D expenditures on biotechnology in the higher education and institute sectors amounted to PPP$ 90.2 million; of this amount PPP$ 37.9 million (42%) was spent on gene technology. The business enterprise sector spent PPP$ 29.3 million on biotechnology R&D.
- Over 70% of biotechnology R&D expenditure in 2003 came from public sources.
- In the higher education and institute sectors combined, the Human biomedicine and biopharmacy field was allocated the most R&D resources in 2003 (PPP$ 31.3 million).
- In the business enterprise sector, the Research & Development sector (Services sector NACE 73) had the highest R&D expenditures on biotechnology (PPP$ 11.2 million), followed by the chemical sector (PPP$ 10.2 million).
- In 2003, 1,440 biotechnology researchers worked in the higher education and institute sector; about half of these were women.

Biotechnology in Poland

- In 2005, the Ministry of Education and Science (formerly the Ministry of Scientific Research and Information Technology) ran a biotechnology firm survey for reference year 2004.
- This survey was the first of its kind and included the OECD definition of biotechnology.

- The survey was mandatory. The survey response rate was 34% and there was no weighting for non-respondents.
- In 2004, there were 13 biotechnology firms in Poland.
- Firms were classified into six areas of application: Bioinformatics, Environment, Health, Agriculture & Food processing, Industrial biotechnology and Non-technical areas of biotechnology.
- In 2004, 5 firms were classified in the Health area, 3 firms were classified in Environment, 2 firms were in Agriculture & Food processing, and the Bioinformatics and Industrial biotechnology areas had 1 firm each.
- In 2004, biotechnology firms spent PPP$ 8.7 million on biotechnology, 54% of which was spent on biotechnology R&D and 36% on biotechnology capital (instruments, equipment, land and buildings).
- In 2004, total business enterprise R&D expenditure by all firms in Poland was PPP$ 807.9 million, 0.58% of which was spent on biotechnology R&D.
- In 2004, the 13 biotechnology firms employed 946 biotechnology persons (headcount data), of these 109 (12%) had biotechnology R&D responsibilities.

Biotechnology in South Africa

- In 2003, the South African Department of Science and Technology (DST) and eGoli Bio, a Life Sciences Incubator, commissioned a National Biotechnology Survey.
- This survey, covering reference year 2002-03, was the first of its kind in South Africa.
- The survey was modeled on the recommendations of the OECD *Ad hoc* Biotechnology Statistics group

and provided both the OECD list-based definition of biotechnology and a modified single definition.

- The survey scope covered: research groups, firms, research projects, and support service groups. Support service groups included venture capital and private equity firms, government and non-governmental organisations and private organisations.
- The survey focused on 'modern' biotechnology.
- The survey had an overall response rate of approximately 72%; results were not extrapolated to reflect the entire population.
- In 2002-03, 106 firms — 47 core and 59 noncore — were involved in biotechnology in South Africa. Traditional biotechnology firms (brewing, food & beverage, wine) were excluded unless they used modern technologies.
- In the company survey, financial data was requested for financial year 2002-03.
- Biotechnology turnover in the business sector was estimated to be at least PPP$ 123 million in 2002-03.
- Biotechnology R&D was estimated to be PPP$ 83.6 million for the business sector in 2002-03.
- Only 43% of all core and non-core biotechnology firms responded to the employment question. Combined, these firms reported having 1,020 employees with biotechnology-related duties in 2002-03. This figure constitutes a lower bound estimate of biotechnology employment.
- Most core biotechnology firms were in the human health sector (18 firms), followed by the support services sector (6 firms).

- Eighty-nine percent of biotechnology firms had less than 50 employees in 2002-03.
- The firms were asked to classify the biotechnology used within six categories: 1st Generation, 2nd Generation, 3rd Generation, Natural Products, Technology Platforms and Support Services. For more information on the methodology undertaken by the South African Department of Science and Technology a study was released in 2004 and is available on line, at: http://www.oecd.org/dataoecd/7/37/36036991.pdf.
- Over half of the core and non-core biotechnology firms in South Africa were using first generation biotechnology through the application of modern technologies.
- In 2002-3, there were 622 research groups undertaking biotechnology-related activities.

These groups were broken down further into 'biotech' (296), 'potential biotech' (205) and 'biotech services' (178).

- In 2002-03, most research groups were in the human health sector followed by the plant sector.
- Combined, biotechnology firms and research groups were responsible for about 154 biotechnology products in 2002-03. The majority of these were human health products (22%) followed by support services products (20%).
- In 2002-03, 911 biotechnology research projects were identified, including projects undertaken by both research groups and firms.
- The dominant sector for research projects was plant biotechnology followed by human health.

Biotechnology in Spain

- The National Statistics Institute (INE) added a module on biotechnology R&D activities to its 2004 R&D and Innovation survey. The survey included the OECD definition of biotechnology.
- The survey is annual and mandatory.
- The response rate to the 2004 survey was 86% and the results were weighted for nonrespondents.
- The survey covers the whole economy, including both the public and private sector.
- According to the 2004 survey, 278 firms, 102 government sector institutes, 40 universities, and 7 private non-profit institutes carried out biotechnology R&D.
- In 2004, total public and private sector R&D expenditure on biotechnology amounted to PPP$ 656.3 million.
- In 2004, the government sector had the largest share of total biotechnology R&D expenditure, with 36% or PPP$ 238.8 million. The higher education sector had the second largest share of total biotechnology R&D expenditure, with 33% or PPP$ 213.8 million, the business enterprise sector was third, with 30% or PPP$ 198.7 million.
- The remaining results focus on the firms performing biotechnology R&D.
- In 2004, 278 firms reported performing biotechnology R&D in Spain, which represented 5% of all firms undertaking R&D.
- In 2004, R&D expenditure on biotechnology by firms amounted to PPP$ 198.7 million, or 3.1% of all

business enterprise R&D expenditure (PPP$ 6,336.1 million).

- In 2004, 83% of R&D expenditure on biotechnology by firms was spent on current expenditures and 17% on capital investments.
- On average, biotechnology firms spent about 36% of their total R&D budget on biotechnology R&D.
- Source of funding data shows that 96% of business enterprise R&D funds were financed from national funds in 2004.
- In all sectors, 9,444 FTEs were active in biotechnology R&D, of which 6,446 (68%) were researchers and the remainder were technicians and assistants. The business sector accounted for 25% of all FTEs active in biotechnology and 20.8% of all researchers.
- In 2004, 2,387 full-time equivalent employees worked on biotechnology R&D in the enterprise sector. Fifty-six percent of these biotechnology R&D FTEs were Researchers (1,340 FTEs).
- Over half of all biotechnology R&D FTEs employed by the business enterprise sector were women (55% or 1,324 women). Fiftyfour percent of all FTE Researchers working on biotechnology R&D were women (724 FTEs).

Biotechnology in Sweden

- Two Swedish agencies collect biotechnology data: Statistics Sweden and VINNOVA, the Swedish Agency for Innovation Systems.

Statistics Sweden

- Statistics Sweden first added a question on biotechnology R&D personnel to its Business Enterprise R&D (BERD) survey in 1997. No definition of biotechnology was included in any of the surveys. Statistics Sweden plans to include the OECD definition of biotechnology in the 2005 BERD survey.

- The BERD survey is run on a biennial basis and is mandatory since 2001.
- The response rate to the most recent BERD survey, for 2003, was 94% and the results were weighted for non-respondents. The survey was limited to firms with 50 or more employees.
- The 2003 BERD survey estimates 1,648 fulltime equivalents (FTEs) working on biotechnology R&D in 58 firms. Seventy-five percent are in the manufacturing sector.
- Statistics Sweden added a question on biotechnology R&D expenditures to its biennial higher education and government sector R&D surveys in 1997. Neither survey included a definition of biotechnology.
- The higher education R&D survey is run on a voluntary basis. The response rate to the 2003 survey was 88% for the economic data and the results were not weighted for non-respondents.
- In the latest higher education sector survey, for 2003, about 1.3% of total R&D expenditure in this sector was devoted to biotechnology.
- 2003 data for government sector biotechnology R&D were not released.

Swedish Agency for Innovation Systems

- In 2004, VINNOVA, in collaboration with regional organisations, undertook a mapping of firms working in the areas of biotechnology, pharmaceuticals and medical technology in Sweden. This updates earlier studies for 1997 and 2000. Relevant firms were identified from regional organisations, government agencies, industrial research institutes, venture capital companies, etc. Firms were categorised according to business field, focus (R&D, production, consultancy or marketing and sales), national or international market focus and other parameters. About 75% of the companies were contacted to make sure that they were accurately categorised.

- The study was released in February 2005 and is available on line (in Swedish), at: http://www.vinnova.se/main.aspx?ID=0B01F8 AE-24D3-44A3-8180-9154F8ABCFA6.
- In 2003, VINNOVA identified 213 biotechnology firms employing 8,632 persons in Sweden. One hundred and fifty-eight of these firms had less than 50 employees (74%).

The study includes biotechnology units and spin-offs from two large pharmaceutical firms active in Sweden: Pharmacia and Astra/Astra- Zeneca. However, the activities of the two parent firms in Sweden were not included. The number of firms was up from 136 firms in 1997, for an average annual growth rate of 7.8%. The average annual growth rate for employees over the same period was 12.4%.

- From 1997 to 2003, the sectors with the greatest rise in the number of firms were 'Biotech tools and supplies' (14.6%) and 'Drug discovery & development' (12.2%).
- From 1997 to 2003, the sectors with the greatest rise in employment were 'Biotech medical technology' (50.8%) and 'Drug discovery & development' (28.2%).
- In 1997, 2000 and 2003, the fields with the largest number of biotechnology firms were, respectively,

'Drug discovery & development' (60 firms in 2003) and 'Biotech tools & supplies' (59 firms in 2003).

- In 2003, the sector that employed the largest number of persons was 'Biotech tools & supplies', with 2,644 employees, representing 31% of total biotechnology employment. In 1997, 'Biotech tools & supplies' represented 37% of total biotechnology employment.

Biotechnology in Switzerland

- The Swiss Federal Statistics Office business enterprise R&D survey is a voluntary survey run every four years. The results of the survey are corrected for non-response and extrapolated to reflect the entire population.
- A question on biotechnology was added to the 2000 business enterprise R&D survey; the question was again included in the 2004 survey, this time with the OECD definition.
- In 2004, the overall response rate was 81%.
- In 2004, 157 firms reported undertaking biotechnology R&D.
- In 2000, intramural biotechnology R&D in the business enterprise sector accounted for 3.8%, or PPP$ 159.1 million, of total business enterprise R&D expenditures. By 2004, this amount had almost trebled to PPP$ 469.3 million and accounted for 6.8% of total business enterprise R&D expenditures.
- Over half of the PPP$ 469.3 million spent on biotechnology R&D in 2004 was spent by the Pharmaceuticals sector (62% or PPP$ 292.9 million). The Research & Development sector spent the second largest amount on biotechnology R&D (23% of total, or PPP$ 108.7 million).

- Firms with 100 or more employees were responsible for 78% of all biotechnology R&D expenditures in 2004.
- In 2004, Small and Medium-Sized Enterprises (SMEs) spent 76% of their resources in the Research & Development sector (PPP$ 77.4 million out of PPP$ 101.5 million).

SMEs are defined by the Swiss Federal Statistics Office as firms with less than 100 employees.

- For firms with over 100 employees, the Pharmaceuticals sector was responsible for the largest share of biotechnology R&D expenditures (79% of all biotechnology R&D spent by large firms or PPP$ 291.9 million out of PPP$ 367.8 million).

Biotechnology in United Kingdom

- In the United Kingdom, several agencies have made efforts to collect data on biotechnology, including: the Department of Trade and Industry (DTI) and the Office of National Statistics (ONS).

Department of Trade and Industry

- DTI commissioned Critical I Limited, a consulting agency, to collect data on biotechnology. To date, three reports have been produced covering different periods: 2001, 2001 to 2002 and 2001 to 2003.
- The scope of the study was limited to "those companies whose primary commercial activity depends on the application of biological organisms, systems or processes, or on the provision of specialist services to facilitate the understanding thereof."
- In 2003, there were 455 firms involved in biotechnology in the United Kingdom. These firms spent PPP$ 2,008.4 million on R&D.

This figure represents total R&D expenditures undertaken by the 455 firms, as no data was collected for biotechnology R&D alone.

- Over half of the firms involved in biotechnology in 2003 were in the human healthcare sector (53% or 239). These firms spent PPP$ 1,746.8 million on R&D, or 87% of total R&D expenditures.
- In 2003, healthcare sector revenues represented 70%, or PPP$ 4,004.2 million, of total revenues (PPP$ 5,758.9 million). Again, these data represent total revenues and not just biotechnology revenues.
- In 2003, the 455 firms employed 22,405 fulltime equivalent employees, 59% of which were working in the healthcare sector.

Office of National Statistics

- The ONS has been collecting data on biotechnology in its annual government R&D survey since 1993. The survey collects data on central government budget provision for R&D expenditure. The surveys provided a definition of biotechnology.
- In 2003, the UK government spent PPP$ 211.8 million on biotechnology R&D, which represented 1.6% of total government R&D.

Biotechnology in U.S.

- In the United States, several agencies have made efforts to collect data on biotechnology, including: the National Science Foundation (NSF) and the Department of Commerce.

National Science Foundation

- The NSF first added a biotechnology question to its annual business enterprise R&D survey in 2001. Firms were asked to estimate the percent of total

R&D expenditures devoted to biotechnology. The surveys provided a definition of biotechnology. As of 2003, the OECD definition of biotechnology was provided.

- The overall response rate to the business R&D survey was 81.4%, but the section that collects data on biotechnology R&D was not mandatory. There was no weighting for nonrespondents on the biotechnology R&D question.
- In 2003, 2,196 firms were undertaking biotechnology R&D, representing 6% of all R&D-active firms. These firms spent PPP$ 14,232 million on biotechnology R&D, or 7% of total business enterprise R&D expenditures.
- The majority of firms undertaking biotechnology R&D in 2003 were in the nonmanufacturing sector: 73% or 1,609 firms.

These firms were responsible for spending 62% of all biotechnology R&D expenditures; the remainder was spent by the manufacturing sector.

- The Scientific and R&D services sector in the non-manufacturing branch (North American Industry Classification System (NAICS) code 5417) had the highest share of biotechnology R&D expenditures: 36% of all biotechnology R&D expenditures. The Pharmaceuticals and medicines sector in the manufacturing branch (NAICS code 3254) had the second largest share, with 32% of total biotechnology R&D expenditures.
- Ninety percent of the firms undertaking biotechnology R&D were Small and Medium- Sized Enterprises (SMEs) with fewer than 250 employees.

However, these SMEs accounted for only PPP$ 3,614 million of biotechnology R&D, or 25% of the total.

Department of Commerce

- In 2002, the Department of Commerce launched the first US biotechnology firm survey, covering 2001.
- This survey did not represent all US firms engaged in biotechnology, and therefore underestimated biotechnology activity in US industries; however, it still provides insight into biotechnology activities.
- In 2001, the survey identified 1,031 firms engaged in biotechnology-related activities in the United States.
- Over half of all biotechnology firms that responded to the employment question (59% or 600 of 1,025) had 50 or fewer employees. Ninety percent of biotechnology firms had 500 or fewer employees.
- In 2001, over three-quarters of biotechnology firms (76% or 780) indicated that their primary or secondary area of activity was human health applications (HH).
- In 2001, 884 firms reported biotechnology R&D expenditures of PPP$ $16.4 billion, representing about 10% of all US industry R&D in that year.

Ninety percent of all biotechnology R&D was spent in the human health application area.

- Biotechnology sales were PPP$ 50.5 billion or 8.9% of total sales by biotechnology firms in 2001.
- In 2001, biotechnology firms employed 1.1 million employees, of which approximately 11% (66,000) had biotechnology-related responsibilities. Over half of the employees with biotechnology-related responsibilities were scientists (55%).

2

Policy for Research in Biotechnology: A Model

Introduction

The USDA portion of the "Proposal for a Coordinated Framework for Regulation of Biotechnology" appeared at 49 FR 50897-50904. As a part of its policy perspective, USDA stated that agriculture and forestry products developed by biotechnology will not differ fundamentally from conventional products and that the existing regulatory framework is adequate to regulate biotechnology. USDA has both research and regulatory responsibilities for biotechnology activities. This document provides significant new information in both areas. Included in this discussion is an explanation of the assignment of responsibilities within USDA for the oversight of USDA funded research and for the regulation of the products of biotechnology.

An understanding of the way in which USDA has divided these responsibilities should prove helpful to those in the private sector seeking review and/or approval of biotechnology applications. A new section III has been added describing USDA's policy for agricultural biotechnology research. USDA is publishing as a companion document, USDA Guidelines for Biotechnology Research that will closely parallel the NIH Guidelines. The USDA guidelines will be issued under the authority of the Food Security Act of 1985 (Pub. L. 99-198). This Act amended section 1404(2) of the National Agriculture Research, Extension, and Teaching Policy Act (NARETPA). The Amendment gave the Secretary of Agriculture responsibility for establishing "appropriate controls with respect to the development and use of the application of biotechnology to agriculture." All USDA funded agriculture biotechnology research or research conducted at an entity receiving USDA funds would be subject to the USDA Guidelines for Biotechnology Research unless the specific research project is supported by and subject to the guidelines or regulations of another Federal agency.

These Guidelines would encompass all phases of agricultural biotechnology research, i.e. (1) Contained laboratory experiments; (2) specialized isolation research (e.g., greenhouse, biotron); and (3) environmental research release (e.g., controlled and segregated field plots). USDA hopes that entities not required to comply with the Guidelines would voluntarily adhere to the requirements. To encourage compliance, USDA proposes to adopt the NIH policy of providing the researchers not required to comply with these Guidelines the opportunity to have their new biotechnology research proposals reviewed by USDA. Those entities covered by the USDA Guidelines for Biotechnology Research would also be required to comply with any applicable statutes such as those set forth in section IV of this document, and any regulatory issues thereunder. The Secretary of Agriculture has established an Office of Agriculture Biotechnology (OAB), which will have primary responsibility for implementing and coordinating the Department's policies and procedures pertaining to all facets of biotechnology. This includes the conduct of laboratory and field research, exprimentation on biotechnology products prior to their commercialization, and all matters of oversight of biotechnology in agriculture. The new office will report to the Assistant Secretary for Science and Education through the authority provided in the amendment to the Food Security Act of 1985. The Assistant Secretary for Science and Education will seek to establish an Agriculture Biotechnology Recombinant DNA Advisory Committee (ABRAC) and shall continue the responsibilities for agriculture formerly handled by the NIH-RAC during the last 10 years. The OAB shall operate in a close parallel manner to the Office of Recombinant DNA Activities (ORDA) of the National Institutes of Health. This includes the responsibility of the ABRAC and the implementation of the USDA Guidelines for Biotechnology Research. The NIH system is well respected both domestically and worldwide, and has achieved a high degree of efficiency in achieving broad confidence in the safety of new biological research conducted under its requirements. The OAB also will serve as a focal point for coordinating a National Biological Impact Assessment Program, which is to evaluate and monitor the potential impacts of biotechnological processes and products on safety and the environment.

Section IV contains USDA's regulatory policy statements for veterinary biological products, plants and plant products, meat and poultry products, and seeds. USDA stated in the December 31, 1984 Notice that while its existing regulatory framework is adequate, it would constantly reevaluate its regulatory position and should additional regulatory measures become necessary, amend its regulations (49 FR 50904). For veterinary biologicals regulated under the Virus-Serum-Toxin Act (VSTA), USDA has identified three categories which may be derived by recombinant DNA techniques or developed from hybridomas. The categories are based on biological characteristics and safety concerns, and are described fully in section IV(A). The first category consists of inactivated recombinant DNA-derived vaccines, bacterins, bacterin-toxoids, virus subunits, or bacterial subunits, as well as monoclonal products. This category presents no new or unusual safety or environmental concerns. The second category includes those products containing live microorganisms that have been modified by the addition or deletion of one or more genes. Such products will be evaluated under current regulatory policies and procedures to assure that the addition or deletion of specific genetic information does not impart increased virulence, pathogenicity, or survival advantages. The third category included products using live vectors to carry recombinant derived foreign genes for immunizing antigens and/or other immune stimulants. Characteristics of safety and transmission must be established fully before questions and concerns dealing with safety to humans, animals, and release into the environment can be answered and before such products can be considered for licensing. Section IV(A) also includes new information about revised USDA review procedures for the importation of cell cultures and hybridomas.

A brief discussion is included about the proposed regulations implementing the provisions of the amendments to the VSTA contained in the Food Security Act of 1985. For organisms and products derived by the techniques of genetic engineering, USDA is proposing new rules to regulate organisms which are plant pests or which there is reason to believe are plant pests. It is USDA's policy to regulate certain genetically engineered organisms if the donor, vector/vector agent, or recipient organism is a member of a group of organisms that are known to contain plant pests, or if based on experience, USDA determines that a genetically engineered organism or product is a plant pest or if USDA has reason to believe that a genetically engineered organism or product is a plant pest. The proposed regulations are summarized in section IV(B). The USDA policy for regulating meat and poultry products and seeds derived through biotechnology remains substantially as stated in the December 31, 1984 Notice, and appears in section IV (C) and (D). A new section (V) has been added describing the scientific review mechanisms to be established by USDA to assit USDA Agencies in biotechnology research and regulatory decision-making. USDA has established a Committee on Biotechnology in Agriculture (CBA) chaired by the Assistant Secretary for Science and Education and the Assistant Secretary for Marketing and Inspection Services. A detailed summary of comments on the December 31, 1984 Notice and USDA responses appears as section VI. The comments are organized to conform to the form of the December 31, 1984 Notice, with general comments and responses on the USDA regulatory philosophy followed by comments and responses on specific aspects of USDA's regulatory structure.

Notices

Three Federal Register notices concerning the Department's biotechnology related activities have been published subsequent to publication of the December 31, 1984 Notice. On July 19, 1985, a document amending the delegations of authority of USDA to assign responsibility for these Research and regulatory activities (7 CFR Part 2) was published in the Federal Register (50 FR 29367-29368). In this document, the Secretary of Agriculture delegated responsibility to the Assistant Secretary for Marketing and Inspection Services to coordinate the development and carrying out of all matters and functions pertaining to the Department's regulation of biotechnology and to act as liaison on all matters and functions pertaining to the regulation to biotechnology between agencies within the Department and between the Department and government and private organizations. These responsibilities were further delegated from the Assistant Secretary for Marketing and Inspection Services to the Administrator of the Animal and Plant Health Inspection Service (APHIS). Also in this document, the Secretary of Agriculture delegated responsibility to the Assistant Secretary for Science and Education to coordinate the development and carrying out of all matters and functions pertaining to agricultural research involving biotechnology conducted or funded by the Department including the development and implementation of guidelines for oversight of research activities, and to act as liaison on all matters and functions pertaining to agricultural research in biotechnology between agencies within the Department and between the Department and other governmental, educational and private organizations. n1

n1 The Assistant Secretary for Science and Education oversees the research activities of the Agricultural Research Service (ARS), the Cooperative State Research Service (CSRS), the Extension Service (ES), and the Office of Grants and Program Systems (OGPS). The Assistant Secretary for Marketing and Inspecting Services oversees the regulatory activities of the Animal and Plant Health Inspection Service (APHIS), which includes Veterinary Services (VS) and Plant Protection and Quarantine (PPQ); the Agricultural Marketing Service (AMS); and the Food Safety and Inspection Service (FSIS). The policies and procedures of these agencies for biotechnology were described in the USDA portion of the coordinated policy statement at 49 FR 50899-50904. On September 23, 1985, USDA's APHIS published a notice which contained its policy statement and requirements for the control and protection of documents that contain confidential business information concerning biotechnology and the veterinary biologics program (50 FR 38561-38563). On November 14, 1985, the Office of Science and Technology

Policy published a notice in the Federal Register announcing the establishment of the Biotechnology Science Coordinating Committee (BSCC) (50 FR 47174-47195). This Committee is to serve as an interagency forum for coordinating science issues related to research and commercial applications of biotechnology. The notice also stated that USDA will establish a Committee on Biotechnology in Agriculture (CBA) to assit in assuring that research and regulatory decisions and based on the best science available.

USDA Research Policy Statement

USDA supports research to promote and protect the general health and welfare of the people of the United States. n2 Research program include: Studies on production of food and agricultural processing and marketing; identity and development of new crop and animal sources of food, fiber, and energy; increased agricultural efficiency and reduction of dependence on petroleum-based products; development of improved management and conservation of soil, water, forest, and range resources. The programs are fulfilled through State, Federal, and private industry cooperative efforts.

n2 See Addendum for Research Legislative Authorities. In the areas of agricultural research relevant to biotechnology, many plant, animal, and microbial alterations have been developed for release through traditional genetic approaches such as mutagenesis and hybridization. In a complementary vein, beneficial introduction of organisms from abroad have established a sound base for research and regulatory oversight. The experience with these bases provide a substantial knowledge base for conducting evaluations of the safety and efficacy of biotechnology processes and products. USDA will evaluate the environmental impacts in the context of individual experiments that encompass the entire range of experimentation from contained facilities to open field testing. As knowledge and experience are gained, broadly applicable procedures and guidelines will be developed. Particular consideration will be given to the stability of engineered changes and the possibility that genetic elements might be transferred from one organism to another. Also important will be the development of data that will enable predictions of which organisms may become established in new ecosystems, and resulting environmental consequences. USDA considers products developed through biotechnological techniques as no different from those products resulting from research using conventional techniques providing appropriate research review is conducted with established protocols. Agricultural biotechnology research activities require appropriate review to avoid untoward effects on human health and the environment. USDA expects to rely on the existing network of scientific expertise in the agriculture research community. Thousands of plant selections, animal breeding lines, and microorganisms are tested annually at sites under varying climatic conditions through the Nation. This network of scientific expertise permits continual, open assessment of agricultural research and products of that research in the field. USDA has broad statutory authority to conduct and support research in wide ranging areas of agriculture. In addition to the authorities described in the matrix of Federal Laws related to biotechnology found in the Federal Register Notice of November 14, 1985 (50 FR 47174-47195) the Food Security Act of 1985 (Section 1404(2) of the National Agriculture Research, Extension, and Teaching Policy Act Amendments of 1985, Pub. L. No. 99-198), made the Secretary of Agriculture responsible for establishing "appropriate controls with respect to the development and use of the application of biotechnology to agriculture." Through this authority, and pursuant to the Delegation of Authority Pertaining to Biotechnology published in the Federal Register on July 19, 1985 (50 FR 29367-68), the Assistant Secretary for Science and Education will complete development of a national system of agricultural biotechnology research oversight in much the same manner that agriculture has been a part for the last 10 years through the NIH-RAC.

The Assistant Secretary for Science and Education has initiated the establishment of the Agriculture Biotechnology and Recombinant DNA Advisory Committee (ABRAC), to be managed through an Office of Agriculture Biotechnology (OAB) which is a parallel to the National Institutes of Health Recombinant DNA Advisory Committee (NIT-RAC) and Office of Recombinant DNA Activities (ORDA). The OAB will serve as the focal point for developing and coordinating USDA policies and activities pertaining to biotechnology research and will perform related interagency and public liaison functions. OAB will also assist in carrying out the responsibilities assigned to the Assistant Secretary for Science and Education, including the development and implementation of policies and procedures, and guidelines for the conduct of laboratory and field research. All federally-funded agriculture biotechnology research or research conducted at an entity receiving USDA funds will be subject to the USDA Guidelines for Biotechnology Research, which are published as a companion document to this policy statement, unless the specific research project is supported by and subject to the guidelines or regulations of another Federal agency. These Guidelines encompass the entire spectrum of degrees of containment in agricultural biotechnology research i.e.: (1) Contained laboratory experiments; (2) specialized isolation research (e.g., greenhouse, biotron); and (3) environmental research agricultural biotechnology release (e.g., controlled and segregated field plots). Research investigators not required to comply with USDA Guidelines will be encouraged to follow these Guidelines. To assure consistency, USDA adopted the model established by the NIH of providing such researchers with the opportunity to have their biotechnology research proposals reviewed as required by the Guidelines. The USDA Guidelines for Biotechnology Research require that research organization use the Institutional Biosafety Committee (IBC) concept as established by NIH. This requirement assures that each research organization and its investigators employ a multidisciplinary team to assit in carrying out their responsibilities under the Guidelines. The IBC's, as described in the Guidelines, would consist of persons with relevant agricultural expertise in areas such as recombinant DNA technology, biological safety, physical containment, and ecology. Requests for review beyond IBC should be sent to the Office of Agriculture Biotechnology (OAB) through the Assistant Secretary of Science and Education, Room 324-A, Administration Bldg., Washington, D.C. 20250. These Guidelines also would require compliance with existing statutes of the USDA involving the movement of regulated organisms that require the issuance of a permit. The movement of microorganism injurious to plants and animals as well as the movement of certain non-indigenous plants and animals would continue to follow long-established procedures for USDA approval. After review, a permit, if needed, may be issued that allows movement. It is the responsibility of the research scientists to obtain that permit. The Assistant Secretary for Science and Education will complete establishment of a National Biological Impact Assessment Program (NBIAP) as indicated in the USDA Guidelines for Biotechnology Research. NBIAP would serve to assist USDA in the evaluation and monitoring of biotechnology research and impact over time. Coordination of NBIAP will be provided through OAB.

USDA Regulatory Policy Statements

The existing USDA regulatory authority for biotechnology was listed in the matrix of the December 31, 1984 Notice at 49 FR 50860-50874 and described in brief at 49 FR 50898-50899. The statutes considered most applicable to biotechnology applications are the Virus-Serum-Toxin Act (VSTA) of 1913 (21 U.S.C. 151-158), the Federal Plant Pest Act (FPPA) of May 23, 1957 (7 U.S.C. 150aa-150jj), the Plant Quarantine Act (PQA) of August 20, 1912 (7 U.S.C. 151-164, 166, 167), the Organic Act of September 21, 1944 (7 U.S.C. 147a), the Federal Noxious Weed Act (FNWA) of 1974 (7 U.S.C. 2801 *et seq.*), the Federal Seed Act (FSA) (7 U.S.C. 551 *et seq.*), the Plant Variety Protection Act (PVPA) (7 U.S.C. 2321 *et seq.*), the Federal Meat Inspection Act (FMIA) (21 U.S.C. 601 *et seq.*), and the Poultry Products Inspection Act (PPIA) (21 U.S.C. 451 *et seq.*).

Veterinary Biological Products

Under the Virus-Serum-Toxin Act of 1913, 21 U.S.C. 151-158, the USDA exercises regulatory authority over all veterinary biologics imported into the United States or shipped or delivered for shipment interstate. Recent amendments contained in the Food Security Act of 1985 have extended this authority to products which are shipped intrastate or exported, and have given the Department additional enforcement mechanisms such as the power to detain and seize products. Under the VSTA, veterinary biologics may not be shipped or delivered for shipment if they are worthless, contaminated, dangerous, or harmful. Veterinary biological products must be prepared in a USDA-licensed establishment under regulations promulgated by the Secretary of Agriculture. Those products which are imported into the United States must be imported under a permit issued by the Secretary. The pertinent regulations for veterinary biologics are found in Title 9 of the Code of

Federal Regulations, Parts 101 through 117. New regulations will be drafted to implement the provisions of the amendments to the VSTA. Such regulations will provide for a more comprehensive regulatory scheme, including seizure and condemnation and detention procedures. They also will establish procedures to be used in the issuance of special licenses and exemptions provided for by the legislative amendments. Veterinary biological products are defined in the governing regulations, 9 CFR 101.2(w) as "all viruses, serums, toxins, and analogous products of natural or synthetic origin, such as diagnostics, antitoxins, vaccines, live microorganisms, killed microorganisms, and the antigenic or immunizing components of microorganisms intended for use in the diagnosis, treatment, or prevention of diseases of animals." Licensing provisions for veterinary biological products and establishments are found in Part 102 of the USDA regulations (9 CFR Part 102). A product license requires the satisfactory completion of various requirements to assure purity, safety, potency, and efficacy of the products. The specific requirements were discussed in the December 31, 1984 Notice at 49 FR 50899. Pursuant to § 103.3 (a) through (g) of the USDA regulations, a person may be authorized to ship unlicensed biological products for the purpose of evaluating experimental products by treating limited numbers of domestic animals if USDA determines that the conditions under which the experiment is to be conducted are adequate to prevent spread of disease and approves the procedures set forth in the request for such authorization (9 CFR 103.3 (a)-(g)). Upon satisfactory completion of all requirements, including review and acceptance of labels, a U.S. Veterinary Biological Product License may be issued.

The application of new biotechnological procedures for the production of veterinary biological products is expanding constantly. For the purposes of licensing, biologics derived by recombinant DNA-techniques or developed from hybridomas, may be classified into three broad categories. This division is based upon the biological characteristics of the new products and the safety concerns they present, and is wholly analogous to the approach used in other veterinary biologics. The first category includes inactivated recombinant DNA-derived vaccines, bacterins, bacterin-toxoids, virus subunits, or bacterial subunits. These nonviable or killed products pose no risk to the environment and present no new or unusual safety concerns. Monoclonal antibody (hybridoma) products used prophylactically, therapeutically, or as components of diagnostic kits also are included in this category. The second category includes those products containing live microorganisms that have been modified by the addition or deletion of one or more genes. Deleted genes may code for virulence, oncogenicity, enzyme activity, or other biochemical functions. Added genes may result in the expression of new immunizing antigens or the production of novel biochemical byproducts such as beta-galactosidase. Precautions must be exercised to assure that this addition or deletion of specific genetic information does not impart increased virulence, pathogenicity, or survival advantages in these organisms which are greater than those found in natural or wild-type forms. Modifications also must not impart undesirable new or increased adherence or invasion factors, colonization properties, or intrahost survival factors. It is important that genes added or deleted do not compromise the safety characteristics of the organisms. In most cases it is expected that they will be improved, and would therefore not pose any new threat to humans, other animal species, or to the environment. The genetic information to be added or deleted must consist of well-characterized DNA segments. Required licensing data may include base pair analysis, sequence information, restriction endonuclease sites, as well as phenotypic characterization of the altered organism. A comparison is also required to be made between the genetically engineered organism and the wild-type form with respect to biochemical pathways, virulence traits, or other factors affecting pathogenicity. The third category includes products using live vectors to carry recombinant-derived foreign genes that code for immunizing antigens and/or other immune stimulants. Live vectors may carry multiple recombinant-derived foreign genes since they can carry large quantities of new genetic information. They also are efficient at infecting and immunizing target animal species. These properties, for example, make vaccinia virus recombinants very popular subjects for vaccine development programs.

Live vectors currently being evaluated by licensees, applicants, and other research organizations include vaccinia, bovine papilloma virus, adenoviruses, Simian Virus-40, and yeasts. Characteristics of safety and transmission must be examined before questions and concerns dealing with safety to humans, animals, and release into the environment can be answered and before such products can be considered for licensing. USDA will continue to avail itself of additional expertise from the Public Health Service "Interagency Group to Monitory Vaccine Development, Production, and Usage." This interagency committee will be utilized to consider potential human health hazards from the use of veterinary biological products and to review issues such as those arising from the potential effect of organisms potentially pathogenic to people or animals. Veterinary biological products prepared using modern biotechnological procedures such as recombinant DNA, chemical synthesis, or hybridoma technology will be treated similarly to products prepared by conventional techniques. The unlimited number and kind of products that may result from these modern biotechnology procedures make it impossible to define all requirements in specific terms. Each product is evaluated individually to determine what will be necessary to establish its purity, safety, potency, and efficacy. Scientific considerations may dictate generic areas of concerns or the use of certain tests for specific situations. Special assays, preferably using in vitro methods, may be required for potency and stability determinations. Additional tests may be required to assure safety, especially when live microorganisms are present in the biological products. USDA is authorized to issue three types of permits for importing biological products into the United States (9 CFR 104.2). A separate United States Veterinary Biological Product permit is required for each shipment of biological product to be imported. Permits are required for imported biological products used for research and evaluation, distribution and sale, or transit shipment only. Requests for application (U.S. Form 14-5) should be submitted to the Veterinary Biologics Staff, Veterinary Services, Animal and Plant Health Inspection Service, 6505 Belcrest Road, Hyattsville, Maryland 20782. To provide guidance to current or prospective manufacturers employing modern biotechnological methods, the following points are presented:

1. *Recombinant DNA-Derived Products.* Genetic information coding for a product of interest and other sequences not indigenous to the host are referred to as foreign DNA. Recombinant DNA technology encompasses the isolation, characterization, and expression of foreign DNA in organisms or vectors. The specific cloned nucleotide segment coding for the desired product or other foreign DNA segments must be defined in data supporting each license application. These data must also include a description of the source of the DNA and the nucleotide sequence. A vector is a cloning vehicle which provides a suitable origin

of replication necessary for production of foreign DNA. Such replicons may be derived from plasmids, bacteriophages or viruses such as vaccina, bovine pipillomavirus, adenoviruses, or SV-40. Production of functional gene products depends on the efficient expression of cloned DNA-vector complexes in suitable host organisms. Tissue culture cells, bacteria, yeasts, and virus cells may be used as hosts for replication of vectors. The mechanisms of transfer, the copy number, and the physical state of the constructed vector inside the host cell, integrated or extrachromosomal, must be described. USDA's licensing procedure for veterinary biological products derived from recombinant DNA involves a careful evaluation of each product on an individual basis to assure purity, safety, potency, and efficacy. Scientific and safety considerations may require specific safeguards and procedures in some situations. The USDA strongly recommends that all applicants establish Institutional Biosafety Committees which follow applicable provisions of the NIH Guidelines for Research Involving Recombinant DNA Molecules. USDA intends to propose guidelines which specifically relate to veterinary biological products. Amendments of the regulations and standards dealing with veterinary biologics will also be considered.

2. *Chemically Synthesized Antigens.* When the product consists of chemically synthesized polypetides, the appropriate amino acid sequences will mimic the antigenic site or epitope found in the native antigen where one exists. Procedures used to increase or prolong an immune response, such as coupling to carrier proteins or addition of adjuvants, must also be described. Immunological data derived from chemically synthesized peptides must be as definitive as those from natural antigens.

3. *Monoclonal Antibody Products.* The specificity and potency of monoclonal antibody will be compared with those

of similar policyclonal antibody products where appropriate. The sensitivity and specificity of monoclonal antibody products used in diagnostic test kits and their potency characteristics when used therapeutically must be similar to conventional antibody. Monoclonal antibody must be derived from Master Cell Stocks which meet the applicable requirements of 9 CFR 113.52. In addition, as is currently required, a description of cell cloning procedures, preparation, and characterization of cell passages must also be provided. The Outline of Production must describe all processes including scale-up, ascites fluid or cell culture supernatant preparation, purification, concentration, and inactivation, Mouse colonies must bc screened to demonstrate freedom from adventitious agents, especially those detected by the mouse antibody production (MAP) test. If the MAP test discloses the presence of adventitious agents, the product shall not be released unless inactivation procedures approved by Veterinary Services have been performed and tests conducted to ensure proper application of the procedures.

4. *Master Seeds.* Bacterial or viral seed stocks used to prepare veterinary biological products must meet established procedures used to certify Master Seeds for biological products. The Master Seed for recombinant DNA-derived products may consist of a plasmid or virus carrying the inserted gene. This constructed plasmid is then introduced into the appropriate eukaryotic or prokaryotic expression system selected for vaccine production. Genomic DNA may also be transfected directly into a variety of mammalian cells. Alternatively, in such cases, the stable transfected cell could be considered as the Master Seed. The establishment of Master Seeds consisting of constructed plasmids or transfected cells requires submission of background information concerning the recombinant DNA procedures used to isolate, purify, and identify genetic

material from one source and the modification used for inserting of this material into a new host. Data from cloning, isolation, proliferation, and selection of genetically unique cells would be retained by licensed applicants. In order to characterize adequately the foreign DNA used to code for a particular antigen, the manufacturer must provide a nucleotide sequence analysis. Tissue culture-propagated cells from vertebrate animals used for vector propagation and antigen production must meet the requirements of 9 CFR 113.51 or 113.52. If a Master Seed has been accepted by Veterinary Services for use in a licensed product, further genetic modifications may be approved with reduced requirements for additional host animal efficacy studies. Each Outline of Production must be prepared in accordance with 9 CFR 114.9. Outlines must include procedures to ensure consistency in production and recovery of specific antigenic material. Recovery procedures must include removal of excessive antibiotic levels (9 CFR 114.10) and undesirable fermentation byproducts such as excessive levels of bacterial endotoxins. Serial release tests for purity, safety, and potency will be required. In addition product characterization tests may be required to demonstrate consistent gene expression.

Organisms and Vectors

Pursuant to the Act of February 2, 1903, (21 U.S.C. 111), and the VSTA, USDA has authority to issue such regulations and take such measures as may be deemed proper to prevent the introduction or dissemination into the United States for the contagion of any contagious, infectious, or communicable disease of animals and/or live poultry from a foreign country into the United States or from one State or territory of the United States or the District of Columbia to another. The importation into the United States or interstate shipment of organisms and vectors is regulated under 9 CFR Part 122. Organisms and vectors are defined in 9 CFR 122.1 as entities which may introduce or disseminate any cantagious or infectious disease of animals. Such substances may not be shipped interstate or imported without a permit. Permit applications must completely describe the substances, intended use, location of the permittee, and safeguards. A number of revised administrative and technical provisions have been instituted to expedite the USDA review and issuance of permits for importation or organisms and vectors which include cell cultures and hybridomas. No animal-origin biological materials, such as cell cultres, monoclonal antibodies, organisms, vectors, or related material, may be imported into the United States without a Veterinary Services (VS) Permit (VS Form 16-3A).

To obtain a permit, an application (VS Form 16-3) should be submitted to: Import-Export Staff, Organisms and Vectors, VS, APHIS, USDA, 6505 Belcrest Road, Hyattsville, MD 20782. This is different from the permit required to import veterinary biologics pursuant to Part 104 of the USDA regulations governing such products (VS Form 14-5 and 14-6). Applicants must also complete the questionnaire entitled "Importation Information" and submit it with their application. Based upon the information submitted by the applicant, a determination will be made if the material to be imported requires safety testing to ensure it is free from livestock pathogens. Safety testing is conducted at the Foreign Animal Disease Diagnostic Laboratory (FADDL), Plum Island, New York. Applicants will be advised if a safety test is required and will be given an estimate of the cost for conducting the test. Applicants desiring to have material safety tested must enter into a Cooperative Trust Fund Agreement with APHIS, VS, and deposit in advance sufficient funds to cover the estimated cost. The

Import-Export Animals and Products Staff will initiate the Cooperative Trust Fund Agreement. In order to expedite the procedure, VS may issue a permit for the material to be shipped to FADDL pending receipt of the funds and Cooperative Trust Fund Agreement. However, the signed Cooperative Agreement, plus the necessary funds, must be received by VS before testing can be scheduled at FADDL. Usually 60 to 90 days is needed for issuing a permit for importing material to Plum Island, New York, the completion of safety tests, and the transfer of the imported material to the applicant. A minimum of four vials, each containing at least 1 million cells from a uniform lot, is required for the safety testing. When the test is completed and a determination made that the imported material is free from livestock pathogens, the remainder of the imported material is released directly to the importer under conditions specified in the permit. If an importer wishes to import cell cultures and/or hybridoma cells on a regular basis, the applicant may enter into a continuous Cooperative Trust Fund Agreement with VS and establish an escrow account to ensure that unnecessary delays will not occur due to insufficient funds.

Each safety test utilizing susceptible host animals usually cost approximately $2,000 to $3,000. Sometimes it is possible to reduce the cost by pooling samples in one host animal test. Scientists at FADDL developed in vitro safety tests to detect certain livestock pathogens resulting in substantial cost savings for importers. The current cost of each in vitro test is approximately $500, depending upon the type of animal disease present in the country of origin as well as the intended use of the imported material. Safety testing may not be required for some cell cultures imported for human diagnostic purposes and research. Examples of material which could enter without safety testing include cultured human bone marrow cells, amniocentesis samples, and cells imported for karyotype analysis. Applications for such cell cultures will be considered individually. Permit applications are evaluated by a new classification scheme that correlates intended use of imported cell cultures with the level of safety testing conducted at FADDL. The following classification of cell cultures is based on intended use and generally indicates the level of safety testing required. Class I Cell cultures to be used for the production of products such as vaccines, hormones, or other biologicals to be used in livestock, poultry, or for commercial distribution. Requirement: These cell cultures must be safety tested at FADDL using susceptible host animals, approved in vitro test, and/or laboratory animals. Class II Cell cultures to be used only for in vitro studies and not to be used in animals other than primates.

Requirement: These cultures may not require safety testing. The material may be sent directly to the importer when no safety testing is required. The permit (VS Form 16-3A) will specify restrictions such as "FOR IN VITRO LABORATORY TESTS: DO NOT INOCULATE INTO LIVESTOCK, BIRDS, OR LABORATORY ANIMALS." Cell cultures imported under permit which do not require a safety test may not be distributed to other laboratories without prior approval from USDA, APHIS, VS. Applications for the distribution of imported material should be submitted to the USDA, APHIS, VS, Import-Export Staff, Organisms and Vectors. When appropriate, a review is conducted by the Administrator's Parent Committee on Organisms and Vectors. Members of this committee have wide expertise in evaluating safety. Clearance may also require testing in high security facilities at the Veterinary Services, FADDL, Plum Island, New York.

Plants and Plant Products

Pursuant to the authority granted by the Federal Plant Pest Act (FPPA) of May 23, 1957, as amended (7 U.S.C. 150 aa through 150 jj), and the Plant Quarantine Act (PQA) of August 20, 1912, as amended (7 U.S.C. 151 through 164, 166, and 167). USDA has regulatory authority over the movement into or within and through the United States of plants, plant products, plant pests, and any product or article which may contain a plant pest at the time of movement. These articles are regulated in order to prevent the introduction, spread, or establishment of plant pests new to or not widely prevalent in the United States. The regulations implementing this statutory authority are found in 7 CFR Parts 300 through 399. "Plant Pest," as defined by statute, means any living stage of any insects, mites, nematodes, slugs, snails, protozoa, or other intertebrate animals, bacteria, fungi, or parasitic plants or reproductive parts thereof, viruses, or any organisms similar to or allied with any of the foregoing, or any infectious substances, which can directly or indirectly injure or cause disease or damage in any plants or parts thereof, or any processed, manufactured, or other products of plants (7 U.S.C. 150aa(c)). "Movement," as defined by statute, means to ship, deposit for transmission in the mail, otherwise offer for shipment, offer for entry, import, receive for transportation, carry, or otherwise transport or move, or allow to be moved, by mail or otherwise (7 U.S.C. 150aa(g)).

The current permit system requirements for the movement into or within and through the United States of plants, plant products, plant pests, and other articles regulated by FPPA and PQA were fully described in the December 31, 1984 Notice at 49 FR 50900-01. The procedures for issuing permits for the movement of plant pests were discussed separately from plants, plant products and other articles which may contain plant pests at 49 FR 50901-02. USDA regulates the importation of noxious weeds through a permit system similar to that established for plant pests. The existing regulations in 7 CFR Part 360 which designate plants as noxious weeds and establish procedures for obtaining an import permit were described at 49 FR 50902.

Regulation of the Introduction of Organisms and Products Altered or Produced Through Genetic Engineering Which are or Which There is Reason to Believe are Plant Pests

The FPPA and PQA are applicable to the movement of plants, plant products, and other articles and plant pests developed through genetic engineering if such plants, plant products, other articles, or plant pests present a risk of plant pest introduction, spread, or establishment. Under the authority granted by the FPPA and PQA, USDA is proposing new regulations which would impose restrictions on the introduction of organisms and products altered or produced through genetic engineering which are plant pests or which there is reason to believe are plant pests. In accordance with the provisions of the FPPA and PQA. USDA must determine the plant pest status of plants, plant products or articles to be moved into or within or through the United States. The evaluation process for determining what safeguards, if any, can be imposed which would allow the movement of the plant pest without risk that the plant pest would be disseminated were described in the December 31, 1984 Notice at 49 FR 50901-02. For genetically engineered material from dissimilar source organisms (inter-generic combinations), the determination may be complex. Information about genetically engineered organisms produced through the use of donor, vector/vector agent and recipient organisms that are from a list of known plant pests is needed in order that such organisms be properly regulated.

During the past year, USDA has received permit

applications to move genetically engineered organisms into or through the United States. USDA is confident that organisms altered through genetic engineering will play a major role in increased plant yield and improved plant quality. However, a genetically engineered organism derived from organisms that are plant pests also presents a risk of plant pest introduction.

The organisms themselves, the cultures in which they are transported, or their packaging may be contaminated with plant pathogens. Genetic alteration may create a plant pest new to and not widespread in the United States. It is necessary, therefore, to establish appropriate safeguards to prevent the introduction of genetically engineered organisms that pose a threat to agriculture. Other genetically engineered organisms that are not plant pests or where there is no reason to believe such organisms are plant pests would not be regulated. New data have to be required in order to properly evaluate permit applications for those organisms which are plant pests or which there is reason to believe are plant pests. A determination was made that additional data requirements would be incorporated into proposed regulations for those genetically engineered organisms which are of concern under the provisions of the FPPA and PQA. USDA is publishing as a companion document in the "proposed rules section" of this issue of the Federal Register its proposed regulations pertaining to organisms and products altered or produced through genetic engineering which are on plant pests or which there is reason to believe are plant pests. The proposed regulations would establish a new part entitled, "Introduction of Organisms and Products Altered or Produced Through Genetic Engineering Which Are Plant Pests or Which There is Reason to Believe are Plant Pests", in Title 7 of the Code of Regulations (7 CFR), pursuant to the authority of the

FFPA, as amended (7 U.S.C. 150aa-150jj) and the PQA, as amended, (7 U.S.C. 151-164, 166, 167). Such proposed regulations would regulate the importation into and movement within and through the United States as well as prevent the release into the environment of certain organisms, or products altered or produced through genetic engineering, which are plant pests or which there is reason to believe are plant pests.

The proposed regulations would restrict the "introduction" of certain organisms and products altered or produced through genetic engineering, referred to as "regulate articles." In this context, "introduction" means to move into the United States, to release into the environment, or to move interstate, or any attempt thereat." "Release into the environment" means "use of a regulated article outside the constraints of physical confinement that are found in a laboratory, contained greenhouse, or fermenter or other contained structure." USDA's proposed regulations, which are designed to prevent the release into the environment of genetically engineered organisms which are plant pests or which there is reason to believe are plant pests are consistent with the legislative intent of the FPPA. The FPPA was enacted in 1957 and was intended as "gap filing" legislation for the purpose of protecting American agriculture against invasion by plant pests and diseases which are new to or not theretofore known to be widely prevalent or distributed within and throughout the United States. The FPPA also provides USDA with authority to regulate insects or pests that might later be found to be injurious to cultivated crops. The release into the environment of a genetically engineered plant pest is tantamount to the introduction of a plant pest which is new to and not theretofore known to be widely prevalent within and throughout the United States and subject to regulation under the FPPA. It should be noted that "regulated article" would be defined as any organism or product altered or produced through genetic engineering, if the donor organism, recipient organism, or vector or vector agent belongs to a group of organisms designated by the proposed regulations as having plant pests or any organism or product which USDA determines is a plant pest or which there is reason to believe is a plant pest. Under USDA's proposed definition, certain microorganisms would be excluded if the recipient microorganism is non-pathogenic, is non-infectious, and otherwise not a plant pest, and resulted from the addition of genetic material that is well characterized and contains only non-coding regulatory regions. Restrictions would be required for regulated articles because they are plant pests, or because USDA has reason to believe they are plant pests. The proposed regulations would require that a person obtain a permit prior to the introduction of a regulated article and would list specific conditions required for the introduction of a regulated article.

The regulated article could be introduced only if all conditions in the proposed regulations as well as all conditions specified on the permit were met. It is important to note that in considering whether a permit can be issued for the introduction of a genetically engineered organism, USDA will perform the same comprehensive analysis that is used in determining whether a permit can be issued for the movement of a "conventional" plant pest. Such asessment shall include an examination of the factors that were discussed in the December 31, 1984, Notice at 49 FR 50901-02 as part of the evaluation process for determining what safeguards can be imposed which would allow the movement of a plant pest without risk of dissemination. These factors are oriented toward an examination of the ecological and environmental effects of a release of the genetically engineered organism or product into the environment. The proposed regulations also contain provisions for a certificate of exemption for those organisms or products altered or produced through genetic engineering that are not subject to the proposed regulations. A person seeking to introduce an exempt article could voluntarily request a certificate of exemption to facilitate the introduction of the organism or product. The proposed regulations provide a list of groups of organisms which are plant pests or contain plant pests. If the donor, vector/ vector agent, or recipient of the genetically engineered organism is derived from an organism on the list of organisms containing plant pests, such genetically engineered organism would be deemed a "regulated article".

As defined in the proposed regulations, a plant pest includes microorganisms such as bacteria and viruses, and thus a "regulated article" may be a microorganism unless it meets the provisions for exclusion. It is important to note that in some instances certain microorganisms will be subject to joint regulation by USDA and EPA. USDA has jurisdication over certain microorganisms under the FPPA and PQA if the microorganisms are a plant pest. EPA would have jurisdiction under the Toxic Substances Control Act (TSCA) if the microorganism is deemed to be a "new" microorganism or under the Federal Insecticide, Fungicide, and Rodenticide Act, as amended (FIFRA) if the microorginism is to be used as a pesticide. Because each Agency has a different statutory mandate, certain jurisdictional overlaps cannot be avoided. However, EPA and USDA will work cooperatively and simultaneously in the evaluation of genetically engineered microorganisms that fall under the jurisdication of both Agencies. To expedite the review of these microorganisms each Agency will appoint contact persons to coordinate the review to ensure data requests are not duplicated. The specifics of which microorganisms will be subject to dual Agency review, or primarily single Agency review, is set forth in the preamble of USDA's proposed regulations being published as a companion document to this policy statement. That document should be consulted for further information. A key to determining whether a genetically engineered organism will be regulated by USDA is the list of organisms containing plant pests that appears in § 340.2 in proposed Part 340. USDA acknowledges that this is not an exhaustive list, and that it does not attempt to list every pest species. Comments are welcome on the list as well as on other parts of the proposed regulations. In order to solicit as many comments as possible on the list and all other parts of the proposed regulations, USDA has scheduled public hearings in Washington, DC and Sacramento, California, during the 60-day comment period. The time and place of the public hearings as well as the address to send written comments is specified in the preamble to the proposed regulations. USDA believes that through the submission of detailed comments and full participation by public and private interests, USDA will be able to promulgate a final regulation that will prevent the introduction and dissemination of genetically engineered organisms which are plant pests or which there is reason to believe are plant pests, yet not impede the development of biotechnology.

Meat and Poultry Products

The Food and Safety Inspection Service (FSIS) is responsible for assuring the safety, wholesomeness, and proper labeling of food products prepared from domestic livestock and poultry. The Federal Meat Inspection Act (FMIA) and the Poultry Products Inspection Act (PPIA) require FSIS to inspect cattle, sheep, swine, goats, equines, poultry, and food products prepared from them which are intended for use as human food to assure that they are wholesome, not adulterated, and properly labeled, marked, and packaged. Inspection under these statutes is mandatory. The cost of inspection, except for overtime and holiday inspection work, is required to be borne by the USDA. Food, animals and animal products, other than those required to be inspected under the FMIA and PPIA, may be inspected under a voluntary, reimbursable inspection program established under the Agricultural Marketing Act of 1946. Within the framework of food safety statutes, FSIS has developed regulations for research on animals that are administered experimental animal drugs, biologics, and pesticides (9 CFR 309.17 and 381.75). These regulations state that no animal used in any research investigation involving an experimental biological product, drug, or chemical shall be eligible for slaughter at an official establishment unless certain conditions are met. These conditions include any of several different ways of demonstrating that the use of such biological product, drug, or chemical will not result in the products of such animals being adulterated.

Products Subject to Review: FSIS anticipates that many food animals which are subject to the new techniques of modern biotechnology will not differ substantially in appearance, behavior, or general health from currently inspected cattle, sheep, swine, goats, equines, and poultry. They would be subject to the same inspection procedures and regulations as tradionally inspected food animals. FSIS is aware that some genetically engineered animals, such as mosaics, chimeras, and some hybrids, may differ substantially from animals that are inspected currently under the FMIA and PPIA. If such animals are ever intended for use as human food and are presented for inspection at an official establishment, a decision would have to be made as to whether such animals were covered under the FMIA or

PPIA, and if not, whether the FMIA and PPIA should be amended to require inspection of such animals and their products.

Implementation of Review Authority. FSIS's approach toward the review of food animals resulting from the techniques of modern biotechnology consists, in general, of two phases. The first, an experimental phase, focuses on the experimental aspects of vector administrative, gene transfer and gene expression. Since artificial vectors used in animal gene transfer may be considered as either animals drugs or animal biologics, their administration to food animals would be covered under the current regulations on animals used for research (9 CFR 309.17 and 381.75). The requirement that an animal carcass intended for use as human food not be adulterated may require that certain phenotypic, biochemical, and microbiological parameters not be exceeded before the animal can be slaughered for human food. Depending on future developments, FSIS may amend the regulations (9 CFR 309.17 and 381.75) to provide further assurance that the products of animals genetically engineered by certain techniques are not adulterated. The second phase would be carried out under existing regulations (9 CFR Parts 301 through 381) and would focus on the commercial development, production, inspection and labeling of food animals and food animal products.

Seeds

The Federal Seed Act (FSA) (7 U.S.C. 1551 *et seq.*) defines USDA regulatory authority over the importation and interstate shipment of agricultural and vegetable seeds. It does not apply to the production or intrastate distribution of seeds or to seeds other than agricultural or vegetable seeds ("agricultrual seeds" are grass, forage, and field crop seeds). The FSA prohibits interstate shipment of seed that contains noxious weed seeds at levels in violation of the laws of the State of destination or in excess of levels allowed by the Secretary of Agriculture. This provision applies primarily to seed adulterated with noxious weed seed. In a few instances, however, States have determined that a particular variety of agricultural or vegetable seed is itself a noxious weed. In these instances, FSA prohibits the interstate shipment of the seed into those States. The FSA also allows the Secretary to prohibit the importation of agricultural and vegetable seed which is adulterated with noxious weed seed or which is unfit for seeding purposes. The authority granted to the Secretary by the FSA to prohibit the interstate shipment or importation of seeds which are found to be detrimental to the agricutural interests of the United States applies to seeds genetically engineered with the modern biotechnology to the same extent as any other seeds.

Scientific Review Mechanisms

The manner in which both regulation and oversight of research in agriculture-related biotechnology evolves and is implemented in the United States will have a direct impact on the competitiveness of U.S. industry in both domestic and world markets. Inconsistent or unnecessary procedures for regulation and research will place the U.S. scientific effort and U.S. producers at a substantial disadvantage. It also is important that safeguards be built into biotechnological research processes, and that releases be based on careful evaluations while further experience is being gained. Therefore, USDA feels that such regulatory and research decisions must be based on the best science available.

While the responsibilities within USDA for biotechnology

reside with the Assistant Secretary for Science and Education and the Assistant Secretary for Marketing and Inspection Services as the delegates of the Secretary of Agriculture, in carrying out their respective responsibilities based on the best science available, they would be able to take advantage of the expertise and perspectives within the Federal Government through a committee to be called the Committee on Biotechnology in Agriculture (CBA). The CBA, to be chaired by these two Assistant Secretaries, will function both as a policy body in the USDA and a bridge between its research and regulating structures.

Committee on Biotechnology in Agriculture

The objectives of the CBA will include:

To provide advice, when requested, on initiatives, proposals, and policy for agriculture-related regulation and research, and assist in the coordination of these activities; To review scientific issues submitted by agencies within the Department; To assist in identifying data gaps for basic research in agricultural biotechnology; To foster public awareness of the scientific issues in biotechnology; To provide Departmental support for participation in the FCCSET BSCC. USDA expects that the CBA also will utilize existing cooperative entities (e.g., other Federal agencies, universities, State regulatory officials, the public sector, and industry) to acquire, when necessary, information for addressing those issues submitted to it. Such entities, when requested, can provide technical support for sound regulatory and research decisions regarding the use of biotechnology in agriculture and foresty. These entities offer a vast scientific resource upon which USDA can draw.

Summary of Comments

USDA received the comments of one hundred-two (102) respondents, one-half of whom commented specifically on the USDA policy statement. Although USDA agencies considered all comments on the coordinated policy proposal, this response is confined to comments on the USDA portion of the notice. The two largest categories of respondents were business and academic, followed closely by associations representing these interests. Comments came in lesser numbers from environmental and public interest groups, individuals, law firms, and foreign governments, as well as the National Institutes of Health Recombinant DNA Advisory Committee (NIH-RAC) and a member of the U.S. Congress The USDA response to the comments follows the form of the original notice, with a discussion of comments on regulatory philosophy followed by a response to comments on the regulatory framework.

Comments on the Nature of Products of Modern Biotechnology: Fourteen respondents stressed their agreement with the USDA statement that "agriculture and forestry products developed by modern biotechnology will not differ fundamentally from conventional products," while six commenters dissented. Three respondents felt that genetic engineering across species barriers did create a potentially different product and the possiblity of unique ecological effects. Concern about the "need for public trust" and public assurance on safety and ethical issues was stressed by three commenters. Seven respondents agreed with USDA that "to date, no unique or safety problems have been associated with products of genetic engineering," but four of the same commenters who view biotechnology products as fundamentally different from conventional products stressed that the potential exists for safety problems with biotechnology applications.

Response: USDA recognizes the importance of ecological effects and the need for developing procedures responsive to public concerns about safety. Although USDA's regulatory philosophy remains as stated, additions to regulatory procedures are being proposed for genetically engineered plants and plant products and veterinary biologics produced by biotechnology (see section IV). The previously discussed delegations of authority within USDA for biotechnology increase the effectiveness of the administration of current and proposed regulatory procedures affecting the products of modern biotechnology. For veterinary biological products, USDA is currently developing additional procedures pursuant to the VSTA, as amended, for evaluating requests to conduct experimental field trials with live vectors containing genetically engineered organisms or to support product license applications. The procedures being developed consider the parental organism and the effect of the gene alteration on the genetic properties of the recipient, especially the survival, reproduction, and dispersal characteristics. A careful analysis of the genetics, biology, and ecology of the wild-type and modified microorganisms will provide as reasonable prediction of the risks which might be associated with use of the altered organisms. USDA is proposing regulations pursuant to the Federal Plant Pest Act (FPPA) and the Plant Quarantine Act (PQA) for regulating the introduction of certain organisms of products thereof altered or produced through biotechnology which are plant pests or may become plant pests. This proposed rule should assist USDA in assessing the ecological effects of the release of such genetically engineered organisms into the environment. Guidelines for oversight of agricultural biotechnology research funded by USDA will be issued under the authority of the Food Security Act of 1985. USDA also is establishing scientific review mechanisms to assist in research and regulatory decisions. These proposed modifications in the procedural framework are described as a part of the final policy statement for veterinary biologics, plants and plant products, research, and scientific review mechanisms.

Comments of the Adequacy of Existing Authority: Thirteen commenters agreed with USDA that its existing regulatory framework is adequate for biotechnology applications, and nine favor the case-by-case approach under existing authority. Five commenters felt that new legislation is or may be needed; two of the five oppose the case-by-case approach.

Response: USDA has examined its statutory authority for regulating biotechnology products and processes, and USDA agencies have processed licensing and permit applications under the existing statutes. The existing authority is considered adequate at this time. Established procedures, with the proposed modifications, can be adapted effectively to handle biotechnology applications. USDA is currently considering genetic engineering applications on a case-by-case basis using existing authority.

Comments on Need for Procedures and Guidelines: Sixteen respondents commented that USDA had not outlined procedures for the review and approval of genetically engineered products. Twelve respondents stressed the need for flexibility, and six requested sunset provisions in USDA biotechnology regulations.

Response: The USDA policy statement of December 31, 1984, did outline procedures currently used for the review and approval of certain genetically engineered products. In considering license applications for genetically engineered veterinary biologics, USDA follows the standards and procedures applicable to all such products found in § § 101-117 of the applicable regulations and standards (9 CFR 101-117). In the December 31, 1984 Notice, USDA offered supplementary guidelines for licensing such products. New procedures are being developed to evaluate production and testing of veterinary biologics derived through use of genetic engineering techniques. The information needed for proper evaluation will depend on the parent organism and the effect of the gene alteration on the genetic properties of the recipient. A paper describing the USDA licensing policy for biologics produced by recombinant DNA technology was presented at the Joint International Association of Biological Standardization/World Health Organization Symposium on "Standardization and Control of Biologics Produced by Recombinant DNA Technology," Geneva, Switzerland, 1983 (published in *Developments in Biological Standardization,* V. 59, pp. 167-173, S. Korgel, Basel, 1985). The paper describes requirements for plasmid/vector characterization and stability, and correlation to conventional Master Seed concepts, as well as methodology which can be used to monitor antigenic expression, concentration, purification, and stability testing during production and recovery.

The movement of genetically engineered products which are plant pests and present a risk of plant pest introduction or spread is regulated by 7 CFR 330.200 implemented pursuant to the FPPA and PQA. The movement of organisms and vectors which may cause desease in animals is regulated under 9 CFR Part 122. USDA realized that the statement left unanswered some questions about the means for review and approval of various genetically engineered products.

The proposed regulations described in section IV(B), implemented under the authority of the FPPA establish permit requirements for the "introduction" of organisms altered or produced by genetic engineering which are or may become plant pests.

The regulations would be flexible because organisms determined not to be plant pests would be exempt, and this category could be expanded in the future to include organisms whose plant pest status is currently uncertain and therefore restricted. It is hoped that the discussion in section IV(B) of this policy statement answers any remaining questions about the review and approval procedures for such genetically engineered products.

Comments on Confidential Business Information (CBI): Six commenters representing business and scientific interests expressed concern about the protection of "confidential business information" in the USDA regulatory process while two public interest groups stressed the "public's right to know."

Response: The USDA regulations implementing the Freedom of Information Act (FOIA) (5 U.S.C. 552) are found in 7 CFR 1.1-1.16. The FOIA provides that Federal agencies must make available to the public all records not specifically exempt from disclosure. Exemptions include "trade secrets and commercial or financial information." (5 U.S.C. 552(b)(4)). On September 23, 1985, USDA's APHIS issued a policy statement on the protection of privileged or confidential information (50 FR 38561-38563). This policy statement establishes requirements for the control and protection of documents received by APHIS that contain privileged or confidential business information concerning biotechnology and the veterinary biologics program. The procedures established conform to the FOIA requirements for both protection and disclosure.

Comments on Use of NIH Guidelines: Four respondents questioned the USDA requirements that manufacturers of veterinary biological products using recombinant DNA technology follow the National Institutes for Health Guidelines for Research Involving Recombinant DNA Molecules (NIH Guidelines). One respondent thought USDA implied that all people "who work with recombinant DNA plants" would be required to comply with the NIH Guidelines, and requested procedural guidelines for industry.

Response: The USDA does not require that the manufacturers of veterinary biological products or plant products of recombinant DNA technology follow the NIH Guidelines. However, USDA strongly recommends that all license applicants for veterinary biologics follow appropriate provisions of the NIH Guidelines, such as those regarding the establishment of an institutional biosafety committee. USDA intends to propose guidelines that will parallel closely the NIH Guidelines, and it intends to recommend strongly that entities not required to follow the USDA guidelines do so voluntarily.

Comments on Importation of Cell-Lines: Three associations representing biotechnology companies requested that USDA take steps to reduce delays in the clearance and testing procedures required for the importation of biotechnology-derived products and cell-lines. On February 12, 1985, the Association of Biotechnology Companies (ABC) delivered a report on USDA importation guarantine issues to the APHIS Parent Committee for Foreign Pathogens and Vectors. This report was an attachment to the ABC comment letter.

Response: The USDA has instituted a number of revised administrative and technical provisions to expedite the issuance of permits for importation of organisms and vectors which include cell cultures and hybridomas. A supplementary questionnaire, designed to insure adequate information on cell cultures and products from recombinant DNA or hybridoma technologies, now accompanies each permit application. Applicants are advised whether or not a safety test is required and a cost estimate is given. Safety testing may be conducted concurrently with the administrative review of the permit application, but only at APHIS' Foreign Animal Disease Diagnostic Laboratory (FADDL) at Plum Island, New York. New test procedures have reduced the cost of safety testing, and the cost per sample can be further reduced by conducting a safety test with pooled samples. Permit applications are evaluated using a new classification scheme that equates intended use of imported cell cultures with the level of safety testing required at FADDL. Class I cell cultures, employed in the preparation of products such as enzymes, vaccines, or hormones for commercial use, are subject to complete safety testing. Class II cell cultures, used only for in vitro studies and not to be used in animals other than primates, are subject to a lesser degree of testing.

Comments on Risk Analysis: Seven respondents discussed the issue of risk assessment or risk/benefit analysis of biotechnology applications. Comments varied from a recommendation that "standard risk assessment methodologies" be adopted by all agencies to a warning against attempting to regulate the "hypothetical and imaginary "potential' dangers" of recombinant DNA techniques.

Response: The National Environmental Policy Act (NEPA) applies to USDA actions. The "APHIS Guidelines Concerning Implementation of NEPA Procedures" (44 FR 50381, August 28, 1979) would be used to make an environmental assessment or environmental impact statement of the effects of a proposed release of a genetically engineered organism regulated pursuant to the VSTA, the FPPA and PQA, and related statutes. A formal risk management procedure based on a wide variety of safety concepts will be used to evaluate systematically proposed releases. The information required by any new regulations promulgated under the FPPA and PQA would be used to prepare the environmental assessment for release of a genetically engineered product which is a plant pest or may become a plant pest.

In normal husbandry and laboratory practices, veterinary biological products normally are not considered to be released into the environment. In the event that a conventionally prepared or recombinant derived product would be considered to be released into the environment, the issuance of a license or import permit would require compliance with procedures being developed and interagency approval. The procedures under development consider the parental organism and the effect on the gene alteration on the genetic properties of the recipient, especially the survival, reproduction and dispersal characteristics. Safety, ethics, and policy issues in agricultural biotechnology research will be overseen by the Committee on Biotechnology in Agriculture (CBA) and such supporting technical advisory groups as may be established by the USDA agencies. Currently, all USDA and USDA-sponsored research involving recombinant DNA must be cleared prior to initiation for compliance with the NIH Guidelines.

Comments on Jurisdiction: The potential for overlapping jurisdiction in the policy notice drew the largest number of comments. Eighteen respondents pointed out that both USDA and EPA propose to regulate agricultural microorganisms. Respondents representing the interests of the veterinary biologics industry contended that a jurisdictional dispute between USDA and FDA delayed the approval of bovine interferon. While generally supporting the concept of the memorandum of understanding (MOU) between USDA and FDA to resolve jurisdictional disputes, one respondent challenged the legality of the MOU, noting that it contains the statement that "animal biological products generally act through a specific immune process," while USDA's current regulations do not restrict its jurisdiction to products operating through such a mechanism of action. Industry respondents also pointed out that the intrastate producer of veterinary biologics is not regulated by USDA. Two firms and one industry association urged prompt Federal oversight action so that States do not act independently to regulate biotechnology products.

Response: USDA agrees that there is the potential for overlapping jurisdiction among the Federal agencies involved in regulating biotechnology products. USDA and EPA representatives have discussed jurisdiction over genetic engineering applications since 1983. USDA and EPA have begun to establish a regulatory procedure for reviewing certain submissions of genetically engineered microorganism applications, a procedure which has resulted in joint consultation on several proposals for release into the environment of organisms altered by genetic engineering.

For veterinary biologics regulated under the VSTA, use of procedures currently under development will increase USDA effectiveness in evaluating biotechnology license and product applications. The MOU between USDA and FDA was published on June 8, 1982, in an attempt to resolve the issue of new products which fall into the questionable definitional area between animal drugs regulated by FDA and animal biologics regulated by USDA. An interpretation by some that the term animal biologics only includes substances that act through a specific immune process has resulted in some confusion. There is nothing in USDA's current regulations or law which restricts its jurisdiction to products acting solely through this mechanism of action, and because of this fact, the memorandum qualifies its reference to specific immune process by the word "generally." Although efforts will be made to clarify the issue further, it should be noted that there appears to be little uncertainty about whether a particular product is a veterinary drug or biologic. The Food Security Act of 1985 contains amendments to the VSTA that extend USDA's jurisdiction to veterinary biologics which are shipped intrastate or exported. The provisions of the amendments are discussed more.

Comments on the National Biological Impact Assessment Program (NBIAP): Seven respondents commented on the NBIAP, the proposal by the National Association of State Universities and Land Grant Colleges (NASULGC) for establishing a program to assess genetically engineered organisms before they are released into the environment. Three commenters — a member of Congress, a spokesperson for a biotechnology firm, and an officer of an environmental organization — posed questions about the proposal. The questions concerned the NBIAP's statutory or regulatory status; its relation to other USDA agency operations and other Federal agency operations; the processes of risk assessment to be used; its adequacy to review an increasing volume of products; and the appropriateness of biohazard committees as vehicles for review of commercial processes and products. Four respondents representing NASULGC institutions endorsed the proposal stating the view that the agricultural research community has the capability to develop guidelines and assess impacts of biotechnology research and commercial products. The major goal of the program was thought to be insuring the safety of society and the environment.

Response: NBIAP is a scientific advisory system that would be available to the Assistant Secretary for Science and Education. By this system the USDA can draw upon the best experience available from scientists in universities, Federal laboratories, and industry to help assess the risks involved in the processes and products from RDNA work in biotechnology. NBIAP shall act in an advisory capacity and is in no direct way a part of the formal approval process. It is available to provide assessment, but is not a mandatory process.

Comments on Definitions, Terms, and Data Requirements: Five respondents recommended changes in the definitions, terms, data requirements or classification used by USDA in the notice. Each recommendation is discussed below. Two respondents commented on the USDA statement of licensing policy for veterinary biologics produced by modern biotechnical methods at 49 FR 50899-50900. Under the heading "1. Recombinant DNA-Derived Products," a manufacturer of veterinary biologics questioned the need to provide the entire nucleotide sequence of a foreign DNA being cloned into a vector. It is USDA's position that in order to characterize adequately the foreign DNA used to code a particular antigen, the manufacturer should provide a nucleotide sequence analysis. The construction of the vector used for expression of the cloned nucleotide sequence also should include source and function of the component parts of the vector, i.e., origin of replication, antibiotic resistance genes, promotor, enhancers, etc. The manufacturer also questioned the data requirement under the heading "2. Chemically Synthesized Antigens" concerning the persistence of the immune response following administration of the synthetic peptide. The USDA feels that a major concern with the use of synthetic peptides is the development persistence of the immune response. USDA does not intend to require more stringent efficacy data than that necessary to support a veterinary biologic license application employing natural antigens. However, immunological data derived from chemically synthesized peptides must be as definitive as the serological response from natural or nonsynthetic antigens. With respect to the next sentence in the policy statement, an individual respondent proposed a change from the term "antibody response" to "immune response." It is true that the term used in the sentence "Procedures used to increase or prolong an antibody response . . ." is somewhat limiting and can create confusion between B-cell and T-cell response. Therefore, the recommendation to replace "antibody response" with the term "immune response" with the term "immune response" is accepted, since both T-cell responses as well as T-cell/B-cell interactions would be included in the statement. On the subject of plants and plant pests, a plant pathologist commented on the references to *Pseudomonas syringae* as plant pathogens under the heading "ice nucleation negative bacteria" at 49 FR 50902. The respondent noted that none of the strains of *Pseudomonas syringae* currently proposed for use are plant pathogens and that it would be more correct to call *P. syringae* plant-associated bacteria, some of which are pathogens. USDA will clarify future references to these organisms as the respondent suggests. According to current practice, and under the proposed FPPA regulations, an applicant for a USDA permit to import or move *Pseudomonas syringae* would be required to submit data to show whether of not the strain was a plant pest.

Addendum — Research Legislative Authorities

The USDA is authorized under its Organic Act (7 U.S.C. 2201 et seq.) and other legislation to conduct and support research in wide ranging areas of agriculture. Examples of such other laws include: The Alcohol Fuels Research (7 U.S.C. 3154); the National Latex Commercialization and and Economic Development Act (7 U.S.C. 178-178n); the Animal Health and Disease Research Act (7 U.S.C. 3195); Special Research Grants (7 U.S.C. 450i(c)); The National Aquaculture Act (16 U.S.C. 2801 et seq.); the Cotton Research and Promotion Act (7 U.S.C. 2101 et seq.); the Potato Research Information Act (7 U.S.C. 2611-2627); the Egg Research and Consumer Information Act (7 U.S.C. 701 et seq.); the Beef Research and Information Act (7 U.S.C. 2901 et seq.); the Wheat and Wheat Foods Research and Nutrition Education Act (7 U.S.C. 3401 et seq.); the Animal Cancer

Research Act (7 U.S.C. 3901 et seq.); the Floral Research and Consumer Information Act (7 U.S.C. 4301 et seq.); and the Forest Research Assistance Act (16 U.S.C. 582a-582a-7).

DEPARTMENT OF LABOR

Background

The Occupational Safety and Health Administration (OSHA) published an announcement of guidelines on occupational safety and health in the Federal Register, Volume 50, Number 71, page 14483, April 12, 1985 with a request for public comment. All comments received supported the statement although the National Institute for Occupational Safety and Health (NIOSH) recommended increased surveillance because their research indicated gaps in current knowledge. Biotechnology is the application of biological systems and organisms to technical and industrial processes. The technologies employed in this area include, but are not limited to:

1. Classical genetic selection and/or breeding for purposes such as developing bakers yeast, conventional fermentation and vaccine development;
2. The direct in vitro modification of genetic material, e.g., recombinant DA or gene splicing; and,
3. Other novel techniques for modifying genetic material of living organisms, e.g., cell fusion and hybridoma technology.

Modern biotechnology is analogous to other conventional industrial processes and has great potential benefit to society and wide application to numerous industries. It is considered by some to have economic potential comparable to the microprocessor industry. Genetic engineering has a wide spectrum of applications of commercial importance, but many such applications are in the early stages of development or have been expressed only as concepts. The Occupational Safety and Health Act of 1970 (OSH Acr) grants the Secretary of Labor broad power to require employers to provide a safe and healthful workplace for their employees. Where other Federal agencies exercise their statutory authority to prescribe or enforce standards or regulations affecting occupational safety or health, OSHA is preempted by section 4(b)(1) of the Act. Section 5(a)(1) of the Act requires employers to furnish their employees with a workplace "free from recognized hazards that are causing or are likely to cause death or serious physical harm." Section 5(a)(2) requires employers to comply with safety and health standards set by the Secretary. The Secretary in establishing standards to deal with toxic materials and harmful physical agents is required by the OSH Act to "set the standard which most adequately assures, to the extent feasible, on the basis of the best available evidence, that no employee will suffer material impairment of health or functional capacity even if such employee has regular exposure to the hazard dealt with by such standard for the period of his working life" (section 8(b)(5)). Under a recent Supreme Court decision permanent standards can be promulgated only upon a finding by the Secretary that the standard is reasonably necessary to remedy a significant risk of material health impairment. Finally, emergency temporary standards may be promulgated only upon a finding that employees are "exposed to grave danger." (section 8(c)(1).)

In view of the statutory criteria briefly outlined above and the currently known hazards from biotechnology processes there does not appear to be a need for new OSHA regulations. Furthermore, the biotechnology processes, whether present in laboratories, pilot projects or industrial plants, usually involve conventional chemicals and processes that are already covered by OSHA regulations. These conventional processes use solvents or products, some of which may be toxic or dangerous to employee health in certain dosages over certain periods of time. The potentially hazardous character of some aspects of biotechnology is primarily from the chemicals used and not the biotechnology products. Therefore, the regulations that effectively regulate chemical exposures will usually ensure that biohazards too will be controlled. However, when a process employing biotechnology alone or in combination with conventional chemicals and technology presents a significant hazard to employees which cannot be dealt with by existing standards or the general duty clause, OSHA will consider regulating in order to protect employees health. Increased industrial hygiene monitoring and medical surveillance will help to assure worker protection. At this time, no new regulations that would specifically cover biohazards are warranted. OSHA endorses the BSCC definitions of "intergeneric (new) organism" and "pathogen" found in the preamble, believing they describe the microorganism appropriate for review when environmental or agricultural applications of microorganisms are contemplated. For contained commercial manufacturing processes, these definitions may also properly exclude from review certain microorganisms of known low risk. OSHA is committed to the policy described in the section entitled "International Aspects" in the Office of Science and Technology Policy General Preamble, published in today's Federal Register.

Guidelines

As stated above, section 5(a) of the OSH Act requires that each employer:

1. Shall furnish to each of his employees employment and a place of employment which are free from

recognized hazards that are causing or are likely to cause death or serious physical harm to his employees;

2. Shall comply with occupational safety and health standards under this Act.

Specific standards which may be applicable include:

— Specific air contaminants (29 CFR Part 1910, Subpart Z).
— Access to employee exposure and medical records (29 CFR 1910.20).
— Hazard communication (29 CFR 1910.1200).
— Exposure to toxic chemicals in laboratories (currently in draft and under development).
— Respiratory protection (29 CFR 1910.134) (currently being updated).
— Safety standards of a general nature, for example, general environmental, walking and working surfaces, fire protection, compressed gases, electrical safety, and material handling and storage contained in 29 CFR Part 1910 Subparts J, D, E and L, H, S and N).

Effective biological safety and health programs have been operative in a variety of laboratories for many years. Motivation and critical judgment are necessary in addition to specific safety and health knowledge to ensure protection of personnel, the public and the environment. All personnel directly involved in biotechnological projects should receive adequate instruction so that the potential biohazards can be understood and appreciated. Emergency plans should be formulated for each project where the chemicals used or biotechnical product product produced pose a potential safety or health hazard. The plans should describe the procedures to be followed if an accident contaminates personnel or workplaces. If a research group is working with a known pathogen for which an effective vaccine is available, employees should be immunized, as appropriate. Before biotechnological work is undertaken, it is imporant that management determine the potential hazards involved and the precautions to be taken. Program and support staff should then be advised of the real and potential hazards. Staff should be instructed and trained in the protection and techniques required to ensure safety and in the procedures for dealing with accidentally created hazards.

Myriad Genetics and Syngenta Complete Rice Genome Map

January 26, 2001 at Salt Lake City

Myriad Genetics, Inc. (Nasdaq: MYGN) announced on that it has completed DNA sequencing of the entire rice genome in collaboration with the Torrey Mesa Research Institute, TMRI, (formerly Novartis Agricultural Discovery Institute, Inc.), a subsidiary of Syngenta. Myriad completed the high-throughput sequencing component of the genome six months ahead of schedule and under budget, triggering a $3 million cash bonus payment to Myriad.

The rice genome map will assist crop researchers and plant breeders in the development of new crop varieties with enhanced characteristics. Additionally, it may serve as a model for other cereal crops such as corn and wheat. In addition to the DNA sequence, the information about the rice genome includes the regulatory DNA sequences controlling gene activity and the location of most of the genes. Furthermore, the collaboration will now be expanded to include the use of Myriad's ProNet® technology to elucidate the rice proteome. The sequencing of the rice genome is an important milestone because it is the first agriculturally important crop to be sequenced. The rice genome is made up of 430 million bases of DNA, representing approximately 50,000 genes, making it the second largest genome of any kind to have been sequenced to date.

Myriad and Syngenta completed the rice genome to a very high degree of accuracy in less than a year and a half from announcement of the project. The rice genome sequence is the most accurate and comprehensive large genome with direct commercial value yet sequenced. The rice genome map is better than 99.5% complete, compared with the human genome map, which still contains a number of large gaps and is estimated to be approximately 90-95% complete.

"We are very proud that the Torrey Mesa Research Institute and Myriad have pioneered this research and that our innovative approach has been successful," said Dr. Steve Briggs, President of TMRI. "Identifying not only the genes, but their functions and how they work, will provide researchers with crucial new knowledge to improve food crops."

This remarkable accomplishment was aided by Myriad's proprietary, high-throughput DNA sequencing facility that averages 40% more sequence per machine than other users of capillary sequencing machines. Myriad's outstanding capability in DNA sequencing pervades the organization, from predictive medicine to proteomics and genome sequencing. The expansion of the collaboration adds the use of the Company's ProNet® technology to determine how protein interactions control the rice plant's growth and development.

"Our Syngenta collaboration highlights many of Myriad's strengths, including our strategy of building multiple sources of near-term revenue generation," said Peter Meldrum, President and Chief Executive Officer of Myriad Genetics, Inc. "The alliance has demonstrated our ability to move away from mere royalties toward 50:50 profit sharing agreements on the use of our technologies."

The rice genome project required extensive use of Myriad's genomic technologies, including its high-throughput DNA sequencing facility that is based upon the latest generation of capillary DNA sequencers. The facility is generating over 12 million bases of DNA sequence daily, a rate that would allow sequencing the entire human genome in a single year. The high throughput DNA sequencing facility is used in most of Myriad's research programs for the delivery of high quality, high throughput DNA sequence.

"Myriad's bioinformatics capabilities made this accomplishment possible in such a short time frame," said Sudhir Sahasrabudhe, Ph.D., Executive Vice President of Research and Development for Myriad Genetics, Inc.Â "The basecalling and sequence assembly software as well as software that constructs individual chromosomes from the segmented data were all developed in-house at Myriad and are available only to Myriad and its collaborators."

Myriad and Syngenta will make the rice genome data available to researchers worldwide through research contracts. In the developing world, where rice is a vital crop, Syngenta will work with local research institutes to explore how this information can best be used to find crop improvements to benefit subsistence farmers.' It is Syngenta's policy to provide such information and technology for subsistence farm-related research without royalties or technology fees. Rice is the largest commercial crop in the world, nourishing over half of the world's population. Knowledge of the rice plant's genes and their function may help plant scientists to introduce new varieties that have enhanced resistance to pests and disease. Additionally, higher yielding varieties could make a major contribution toward the elimination of world hunger.

To access a live Webcast to discuss this scientific breakthrough, please log onto http://www.syngenta.com/, and click on the teleconference icon, at 10:45am eastern time today, for a presentation hosted by Myriad Genetics and Syngenta.

Syngenta is the world's leading agribusiness. The company ranks first in crop protection, and third in the high-value commercial seeds market. Proforma sales in 1999 were approximately US $7 billion. Syngenta employs more than 20,000 people in over 50 countries. The company is committed to sustainable agriculture through innovative Research and Technology. Formed in November 2000 by the merger of Novartis Agribusiness and Zeneca Agrochemicals, Syngenta is listed on the Swiss stock exchange, and in London, New York and Stockholm. Further information is available at http://www.syngenta.com/.

Myriad Genetics, Inc. is a biopharmaceutical company focused on the development of novel therapeutic products derived from its proprietary genomic and proteomic technologies. The Company has established two wholly owned subsidiaries. Myriad Pharmaceuticals, Inc. develops and intends to market therapeutic compounds, and Myriad Genetic Laboratories, Inc. develops and markets proprietary predictive medicine and personalized medicine products. The Company has established strategic alliances with Bayer, Eli Lilly, Hitachi, Novartis, Pharmacia, Roche, Schering AG, Schering-Plough and Syngenta.

3

Mutant Germplasm Charcterization: A Global Perspective

Introduction

Molecular markers have already played a major role in the genetic characterization and improvement of many crop species. They have also contributed to and greatly expanded our abilities to assess biodiversity, reconstruct accurate phylogenetic relationships, and understand the structure, evolution and interaction of plant and microbial populations.

Molecular markers systems reveal variation in genomic DNA sequence. The first generation of molecular markers, RFLP, were based on DNA-DNA hybridisation and were slow and expensive. The invention of the polymerase chain reaction (PCR) to amplify short segments of DNA gave rise to a second generation of faster and less expensive PCR-based markers, which are the main focus of this review. It is already clear, however, that the technology, and future editions of this publication, will continue to move on as new detection systems are developed in the search for evermore efficient and cost effective markers for the breeders of the 21st century.

Restriction fragment length polymorphisms (RFLPs) were the first generation of hybridization-based markers with substantial impact in agricultural biotechnology. These were subsequently followed by amplification-based technologies derived from the polymerase chain reaction (PCR). Notably, the use of arbitrary oligonucleotide primers in the amplification reaction facilitated the study of previously uncharacterized genomes.

The most common of these PCR-based DNA techniques are Amplified Fragment Length Polymorphism (AFLP), Simple Sequence Repeat (SSR), Inter-Simple Sequence Repeat (ISSR) and Randomly Amplified Polymorphic DNA (RAPD). Other techniques have also been developed and are widely used in genetic analysis.

Molecular markers are being used extensively to investigate the genetic basis of agronomic traits and to facilitate the transfer and accumulation of desirable traits between breeding lines. A number of techniques have been particularly useful for genetic analysis.

For example, collections of RFLP probes have been very versatile and important for the generation of genetic maps, construction of physical maps, the establishment of syntenic relationships between genomes, and marker assisted breeding. Numerous examples of specific genes that have been identified as tightly linked to RFLP markers are available for the improvement of specific agronomic traits in almost all major crops. Specific examples include viral, fungal and bacterial resistance genes in maize, wheat, barley, rice, tomatoes and potatoes. Additional examples include insect resistance genes in maize, wheat and rice as well as drought and salt tolerance in sorghum. These markers often used in conjunction with bulked segregant analysis and detailed genetic maps, provide a very efficient method of characterizing and locating natural and induced mutated alleles at genes controling interesting agricultural traits. Markers have also been used to identify the genes underlying quantitative variation for height, maturity, disease resistance and yield in virtually all major crops. In particular, the PCR-based techniques have been useful in the assessment of biodiversity, the study of plant and pathogen populations and their interactions; and identification of plant varieties and cultivars. Amplified DNA techniques have produced sequence-tagged sites that serve as landmarks for genetic and physical mapping. It is envisioned that emerging oligonucleotide-based technologies derived from the use of hybridization arrays, the so-called DNA chips and oligonucleotide arrays, will become important in future genomic studies. However, many of these are still under development, are proprietary, or require the use of expensive equipment, and are therefore not yet suitable nor cost-effective for adequate transfer to developing countries. Clearly, the initial transfer of technology has only involved a selected group of techniques that are well established and/or seem to have a broad application (e.g., RFLP, SSR, ISSR, AFLP, RAPD, IRAP and REMAP and SNPs).

However, techniques are continuously changing and evolving, so technology transfer needs to keep pace with current developments in genomics.

New developments in genomic research have given access to an enormous amount of sequence information as well as new insights on the function and interaction of genes and the evolution of functional domains, chromosomes and genomes. In this context, functional and comparative genomics can help in comparative genetic mapping and linkage analysis of useful agricultural traits. Future DNA marker techniques, such as the use of oligonucleotide arrays, are likely to be sequence based. Comparative analyses of sequence information in the growing databases now publicly available on the World Wide Web will be an invaluable resource for genetic characterisation of exotic crop germplasm.

This will have an increasingly important role in prospection and conservation endeavours.

The following pages in this section present, in very synthetic and basic form, information on the more widely used marker systems; their implementation, application suitability, critical requirements and general comparison of different marker systems. The information is general in nature as many of these techniques are flexible and can be adjusted to suit various requirements.

A Case Study on Cellular Therapy for Patients with Heart Disease

Congestive heart failure remains a major public health problem, with recent estimates indicating that end-stage heart failure with two-year mortality rates of 70-80% affects over 60,000 patients in the US each year. In Western societies heart failure is primarily the consequence of previous myocardial infarction. As new modalities have emerged which have enabled significant reduction in early mortality from acute myocardial infarction, affecting over 1 million new patients in the US annually, there has been a paradoxical increase in the incidence of post-infarction heart failure among the survivors. Current therapy of heart failure is limited to the treatment of already established disease and is predominantly pharmacological in nature, aiming primarily to inhibit the neurohormonal axis that results in excessive cardiac activation through angiotensin- or norepinephrine-dependent pathways. For patients with end-stage heart failure treatment options are extremely limited, with less than 3000 being offered cardiac transplants annually due to the severely limited supply of donor organs, and implantable left ventricular assist devices (LVADs) being expensive, not proven for long-term use, and associated with significant complications. Clearly, development of approaches that prevent heart failure after myocardial infarction would be preferable to those that simply ameliorate or treat already established disease.

Heart Failure After Myocardial Infarction Results From Progressive Ventricular Remodelling. Heart failure after myocardial infarction occurs as a result of a process termed myocardial remodelling. This process is characterized by myocyte apoptosis, cardiomyocyte replacement by fibrous tissue deposition in the ventricular wall, progressive expansion of the initial infarct area and dilation of the left ventricular lumen. Another integral component of the remodelling process appears to be the development of neoangiogenesis within the myocardial infarct scar, a process requiring activation of latent collagenase and other proteinases. Under normal circumstances, the contribution of neoangiogenesis to the infarct bed capillary network is insufficient to keep pace with the tissue growth required for contractile compensation and is unable to support the greater demands of the hypertrophied, but viable, myocardium. The relative lack of oxygen and nutrients to the hypertrophied myocytes may be an important etiological factor in the death of otherwise viable myocardium, resulting in progressive infarct extension and fibrous replacement.

Since late reperfusion of the infarct vascular bed in both humans and experimental animal models significantly benefits ventricular remodelling and survival, we have postulated that methods to successfully augment vascular bed neovascularization might improve cardiac function by preventing loss of hypertrophied, but otherwise viable, cardiac myocytes.

Inability of Damaged Myocardium to Undergo Repair Due to Cell Cycle Arrest of Adult Cardiomyocytes. Cardiomyocytes undergo terminal differentiation soon after birth, and are thought by most investigators to irreversibly withdraw from the cell cycle. Analysis of cardiac myocyte growth during early mammalian development indicates that cardiac myocyte DNA synthesis occurs primarily in utero, with proliferating cells decreasing from 33% at mid-gestation to 2% at birth. While ventricular karyokinesis and cytokinesis are coupled during fetal growth, resulting in increases in mononucleated cardiac myocytes, karyokinesis occurs in the absence of cytokinesis for a transient period during the post-natal period, resulting in binucleation of ventricular myocytes without an overall increase in cell number. A similar dissociation between karyokinesis and cytokinesis characterizes the primary adult mammalian cardiac response to ischemia, resulting in myocyte hypertrophy and increase in nuclear ploidy rather than myocyte hyperplasia. Moreover, in parallel with an inability to progress through cell cycle, ischemic adult cardiomyocytes undergo a high degree of apoptosis.

When cells proliferate, the mitotic cycle progression is tightly regulated by an intricate network of positive and negative signals. Progress from one phase of the cell cycle to the next is controlled by the transduction of mitogenic signals to cyclically expressed proteins known as cyclins and subsequent activation or inactivation of several members of a conserved family of serine/threonine protein kinases known as the cyclin-dependent kinases (cdks). Growth arrest observed with such diverse processes as DNA damage, terminal differentiation, and replicative senescence is due to negative regulation of cell cycle progression by two functionally distinct families of Cdk inhibitors, the Ink4 and Cip/Kip families. The cell cycle inhibitory activity of p21Cip1/WAF1 is intimately correlated with its nuclear localization and participation in quaternary complexes of cell cycle regulators by binding to G1 cyclin-CDK through its N-terminal domain and to proliferating cell nuclear antigen (PCNA) through its C-terminal domain. The latter interaction blocks the ability of PCNA to activate DNA polymerase, the principal replicative DNA polymerase. For a growth-arrested cell to subsequently enter an apoptotic pathway requires signals provided by specific apoptotic stimuli in concert with cell-cycle regulators. For example, caspase-mediated cleavage of p21, together with upregulation of cyclin A–associated cdk2 activity, have been shown to be critical steps for induction of cellular apoptosis by either deprivation of growth factors or hypoxia of cardiomyocytes.

Throughout life, a mixture of young and old cells is present in the normal myocardium. Although most myocytes seem to be terminally differentiated, there is a fraction of younger myocytes (15–20%) that retains the capacity to replicate. Moreover, recent observations have suggested that some human ventricular cardiomyocytes also have the capacity to proliferate and regenerate in response to ischemic injury. The dividing myocytes can be identified on the basis of immunohistochemical staining of proliferating nuclear structures such as Ki67 and cell surface expression of specific surface markers, including c-kit (CD117).

Whether these cells are derived from a resident pool of cardiomyocyte stem cells or are derived from a renewable source of circulating bone marrow-derived stem cells that home to the damaged myocardium remains to be determined. More importantly, the signals required for homing, *in situ* expansion and differentiation of these cells are, at present, unknown. Gaining an understanding of these issues would open the possibility of manipulating the biology of endogenous cardiomyocytes in order to augment the healing process after myocardial ischemia.

Strategies For The Use Of Cellular Therapy To Improve Myocardial Function. Replacement and regeneration of functional cardiac muscle after an ischemic insult to the heart could be achieved by either stimulating proliferation of endogenous mature cardiomyocytes or resident cardiac stem cells, or by implanting exogenous donor-derived or allogeneic cardiomyocytes. The newly formed cardiomyocytes must integrate precisely into the existing myocardial wall in order to augment contractile function of the residual myocardium in a synchronized manner and avoid alterations in the electrical conduction and syncytial contraction of the heart, potentially resulting in life-threatening consequences. In addition, whatever the source of the cells used, it is likely that concurrent myocardial revascularization must also occur in order to ensure viability of the repaired region and prevent further scar tissue formation. The following section discusses various methods of using cellular therapies to replace damaged myocardium or re-initiate mitosis in mature endogenous cardiomyocytes, including transplanted bone marrow-derived cardiomyocyte or endothelial precursors, fetal cardiomyocytes, and skeletal myoblasts.

Potential Role For Bone Marrow-Derived Or Embryonic Cardiomyocyte Lineage Stem Cells In Myocardial Repair/ Regeneration. Over the past several years, a number of studies have suggested that stem cells can be used to generate cardiomyocytes *ex vivo* for potential use in a range of cardiovascular diseases. Multipotent bone marrow-derived mesenchymal stem cells have been identified in adult murine and human bone marrow functionally by their ability to differentiate to lineages of diverse mesenchymal tissues, including bone, cartilage, fat, tendon, and both skeletal and cardiac muscle, and phenotypically by their expression of specific surface markers and lack of hematopoietic lineage markers such as CD34 or CD45. It is well established that murine embryonic stem (ES) cells can give rise to cardiomyocytes in vitro and in vivo. Recently, Kehat et al.

were able to demonstrate that human embryonic stem cells can also differentiate *in vitro* into cells with characteristics of cardiomyocytes. However, there are striking differences in the human and murine stem cell models, and this needs to be taken into account when extrapolating results of mouse experiments to the human condition. For example, human ES cells have a very low efficiency of differentiation to cardiomyocytes compared with murine ES cells, and a considerably slower time course (a median of 11 days vs 2 days). Whether these differences reflect true variations between species, or differences in the experimental protocols, remains to be determined.

Irrespective whether the cardiomyocyte lineage stem cell precursors are obtained from adult bone marrow or embryonic sources, the newly generated cardiomyocytes appear to resemble normal cardiomyocytes in terms of phenotypic properties, such as expression of actinin, desmin and troponin I, and function, including positive and negative chronotropic regulation of contractility by pharmacological agents and production of vasoactive factors such as atrial and brain natriuretic peptides. However, *in vivo* evidence for functional cardiac improvement following transplantation of adult bone marrow-derived or ES-derived cardiomyocytes has been exceedingly difficult to show to date. In part this may be because the signals required for cardiomyocyte differentiation and functional regulation are complex and poorly understood. For example, phenotypic and functional differentiation of mesenchymal stem cells to cardiomyocyte lineage cells *in vitro* requires culture with exogenously added 5-azacytidine. Alternatively, the poor functional data obtained to date may reflect immune-mediated rejection of cells which have been modified during the *ex vivo* culture process or poor viability due to the lack of a sufficient vascular supply to the engrafted cells (see below).

Potential Role for Autologous Skeletal Myoblasts in Myocardial Repair. An alternative approach to replacing damaged myocardium involves the use of autologous skeletal myoblasts. The procedure involves harvesting a patient's skeletal muscle cells, expanding the cells in a laboratory, and re-injecting the cells into the patient's heart.

Perceived advantages of the approach include ease of access to the cellular source, the fact that immunosuppression is not needed, and the lack of ethical dilemmas associated with the use of allogeneic or embryonic cells. It has also been argued that using relatively ischemia-resistant skeletal myoblasts rather than cardiomyocytes might enable higher levels of cell engraftment and survival in infarcted regions of the heart, where cardiomyocytes would probably perish.

Successful engraftment of autologous skeletal myoblasts into injured myocardium has been reported in multiple animal models of cardiac injury. These studies have demonstrated survival and engraftment of myoblasts into infarcted or necrotic hearts, differentiation of the myoblasts into striated cells within the damaged myocardium, and improved myocardial functional performance. Other studies have shown that the survival of transplanted myoblasts can be improved by heat shock pretreatment, and have confirmed that the benefits of skeletal myoblast transfer are additive with those of conventional therapies, such as angiotensin-converting enzyme inhibition. More recently, the procedure has been reported anecdotally to result in improved myocardial function in humans. On the basis of these preliminary results, clinical trials have begun both in Europe and in the United States. In addition to demonstrating functional improvement in large, prospective series, questions that remain to be addressed include whether the skeletal myoblasts can make meaningful electromechanical connections to the surrounding endogenous cardiomyocytes through gap junctions, whether the cellular mass will contract in concert, and whether electrical impulses will be transmitted to the myoblast tissue without inducing significant tachyarrhythmias.

Poor Survival Of Cells Transplanted Into Damaged Myocardium After Ex Vivo Culture. A major limitation to successful cellular therapy in animal models of myocardial damage has been the inability of the introduced donor cells to survive in their host environment, whether such transplants have been congenic (analogous to the autologous scenario in humans) or allogeneic. It has become clear that a major impediment to survival of the implanted cells is the alteration of their immunogenic character by prolonged *ex vivo* culture conditions. For example, whereas myocardial implantation of skeletal muscle in the absence of tissue culture does not induce any adverse immune response and results in grafts showing excellent survival for up to a year, injection of cultured isolated (congenic) myoblasts results in a massive and rapid necrosis of donor myoblasts, with over 90% dead within the first hour after injection. This rapid myoblast death appears to be mediated by host natural killer (NK) cells which respond to immunogenic antigens on the transplanted myoblasts altered by exposure to tissue culture conditions. It seems likely that a similar mechanism of host NK cell-mediated rejection will apply also to transplanted ES-derived, cultured cardiomyocytes since massive death of injected donor cells is recognized as a major problem with transplanted cardiomyocytes, especially in the inflammatory conditions that follow infarction. In this regard, the report that cultured mesenchymal stem cells obtained from adult human bone marrow are not rejected on transfer to other species is intriguing, needs confirmation in humans, and requires detailed investigation into possible tolerogenic mechanisms.

Concomitant Induction Of Vascular Structures Augments Survival and Function Of Cardiomyocyte Precursors. An additional explanation for the poor survival of transplanted cardiomyocytes or skeletal myoblasts may be that viability and prolonged function of transplanted cells requires an augmented vascular supply. Whereas many transplanted cardiomyocytes die by apoptosis, cultured cardiomyocytes that incorporate more vascular structures in vivo demonstrate significantly greater survival. Moreover, in situations where transplanted cardiomyocyte precursors contained an admixture of cells also giving rise to vascular structures, survival and function of the newly formed cardiomyocytes has been significantly augmented.

In one study, direct injection of whole rat bone marrow into a cryo-damaged heart resulted in neovascularization, cardiomyocyte regeneration and functional improvement. More recently, systemic delivery of highly purified bone marrow-derived hematopoietic stem cells in lethally irradiated mice contributed to the formation of both endothelial cells and long-lived cardiomyocytes in ischemic hearts.

Most strikingly, significant improvement in cardiac function of mice who had previously undergone LAD ligation was demonstrated after direct myocardial injection of syngeneic bone marrow-derived stem cells, defined on the basis of c-kit (CD117) expression. This population of cells contains a mixture of cellular elements in addition to cardiomyocyte precursors, including CD117-positive endothelial progenitors (see below). These cells were found to proliferate and differentiate into myocytes, smooth muscle cells and endothelial cells, resulting in the partial regeneration of the destroyed myocardium and prevention of ventricular scarring. Together, these findings raise the intriguing possibility that for long-term *in vivo* viability and functional integrity of stem cell-derived cardiomyocytes it may be necessary to induce neovascularization by coadministration of endothelial cell progenitors (see below).

Endothelial Precursors And Formation Of Vascular Structures During Embryogenesis. In order to develop successful methods for inducing neovascularization of the adult heart, one needs to understand the process of definitive vascular network formation during embryogenesis. In the pre-natal period, hemangioblasts derived from the human ventral aorta give rise to cellular elements involved in both vasculogenesis, or formation of the primitive capillary network, and hematopoiesis. In addition to hematopoietic lineage markers, embryonic hemangioblasts are characterized by expression of the vascular endothelial cell growth factor receptor-2, VEGFR-2, and have high proliferative potential with blast colony formation in response to VEGF. Under the regulatory influence of various transcriptional and differentiation factors, embryonic hemangioblasts mature, migrate and differentiate to become endothelial lining cells and create the primitive vasculogenic network. The differentiation of embryonic hemangioblasts to pluripotent stem cells and to endothelial precursors appears to be related to co-expression of the GATA-2 transcription factor, since GATA-2 knockout embryonic stem cells have a complete block in definitive hematopoiesis and seeding of the fetal liver and bone marrow. Moreover, the earliest precursor of both hematopoietic and endothelial cell lineage to have diverged from embryonic ventral endothelium has been shown to express VEGF receptors as well as GATA-2 and alpha4-integrins. Subsequent to capillary tube formation, the newly-created vasculogenic vessels undergo sprouting, tapering, remodelling, and regression under the direction of VEGF, angiopoietins, and other factors, a process termed angiogenesis. The final component required for definitive vascular network formation to sustain embryonic organogenesis is influx of mesenchymal lineage cells to form the vascular supporting structures such as smooth muscle cells and pericytes.

Characterization of Endothelial Progenitors, or Angioblasts, in Human Adult Bone Marrow. In studies using various animal models of peripheral ischemia a number of groups have shown the potential of adult bone marrow-derived elements to induce neovascularization of ischemic tissues. In the most successful of these, bone marrow-derived cells injected directly into the thighs of rats who had undergone ligation of the left femoral artery and vein induced neovascularization and augmented blood flow in the ischemic limb as documented by laser doppler and immunohistochemical analyses.

Although the nature of the bone marrow-derived endothelial progenitors was not precisely identified in these studies, the cumulative reports indicated that this site may be an important source of endothelial progenitors which could be useful for augmenting collateral vessel growth in ischemic tissues, a process termed therapeutic angiogenesis.

In more recent studies, our group has identified such endothelial progenitors in human adult bone marrow. By employing both *in vitro* and *in vivo* experimental models we have sought to precisely identify the surface characteristics and biological properties of these bone marrow-derived endothelial progenitor cells. Following G-CSF treatment, mobilized mononuclear cells were harvested and CD34+ cells were separated using anti-CD34 mAb coupled to magnetic beads. 90-95% of CD34+ cells co-expressed the hematopoietic lineage marker CD45, 60-80% co-expressed the stem cell factor receptor CD117, and <1% co-expressed the monocyte/macrophage lineage marker CD14. By quadruple parameter analysis, the VEGFR-2 positive cells within the CD34+CD117 bright subset displayed phenotypic characteristics of endothelial progenitors, including co-expression of Tie-2, as well as AC133, but not markers of mature endothelium such as ecNOS, vWF, E-selectin, and ICAM. Sorting CD34+ cells on the basis of CD117 bright or dim expression demonstrated that GATA-2 mRNA and protein levels were significaantly higher in the population, indicating that human adult bone marrow contains cells with an angioblast-like phenotype.

Since the frequency of circulating endothelial cell precursors in animal models has been shown to be increased by cither VEGF or regional ischemia, phenotypically-defined angioblasts were examined for proliferative responses to VEGF and to factors in ischemic serum. However, culture on fibronectin with endothelial growth medium resulted in outgrowth of monolayers with endothelial morphology and functional and phenotypic features characteristic of endothelial cells, including uniform uptake of acetylated LDL, and co-expression of CD34, factor VIII, and eNOS. Thus, G-CSF treatment of adult humans mobilizes into the peripheral circulation a bone-marrow derived population with phenotypic and functional characteristics of embryonic angioblasts, as defined by specific surface phenotype, high proliferative responses to VEGF and cytokines in ischemic serum, and ability differentiate into endothelial cells by culture in medium enriched with endothelial growth factors.

Human Angioblasts Induce Neovascularization Of The Myocardial Infarct Zone. Intravenous injection of freshly-obtained human angioblasts into athymic nude rats who had undergone ligation of the left anterior descending (LAD) coronary artery resulted in infarct zone infiltration within 48 hours. Few human cells were detected in unaffected areas of hearts with regional infarcts or in myocardium of sham-operated rats. Histologic examination at two weeks postinfarction revealed that injection of human angioblasts was accompanied by a significant increase in infarct zone microvascularity, cellularity, and numbers of capillaries, and by reduction in matrix deposition and fibrosis in comparison to controls.

Neovascularization was significantly increased within both the infarct zone and in the peri-infarct rim in rats receiving angioblasts compared with controls receiving saline or other cells which infiltrated the heart (e.g. CD34- cells or saphenous vein endothelial cells, SVEC). The neovascularization induced by human angioblasts was due to both an increase in capillaries of human origin as well as of rat origin, as defined by monoclonal antibodies with specificity for human or rat CD31 endothelial markers. Capillaries of human origin, defined by co-expression of DiI fluorescence and human CD31, but not rat CD31, accounted for 20-25% of the total myocardial capillary vasculature, and was located exclusively within the central infarct zone of collagen deposition. In contrast, capillaries of rat origin, as determined by expression of rat, but not human, CD31, demonstrated a distinctively different pattern of localization, being absent within the central zone of collagen deposition and abundant both at the peri-infarct rim between the region of collagen deposition and myocytes, and between myocytes.

Human Angioblasts Protect Hypertrophied Endogenous Cardiomyocytes Against Apoptosis. By concomitantly staining rat tissues for the myocyte-specific marker desmin and performing DNA end-labeling using the TUNEL technique, temporal examinations demonstrated that the infarct zone neovascularization induced by injection of human angioblasts prevented an eccentrically-extending pro-apoptotic process evident in saline controls. Thus, at two weeks post-infarction, myocardial tissue of LAD-ligated rats who received saline demonstrated 6-fold higher numbers of apoptotic myocytes compared with that from rats receiving intravenous injections of human angioblasts. Moreover, these myocytes had distorted appearance and irregular shape. In contrast, myocytes from LADligated rats who received human angioblasts had regular, oval shape, and were significantly larger than myocytes from control rats.

Human Angioblasts Induce Sustained Regeneration/ Proliferation of Endogenous Cardiomyocytes. In addition to protection of hypertrophied myocytes against apoptosis, human angioblastdependent neovascularization resulted in a striking induction of regeneration/proliferation of endogenous rat cardiomyocytes at the peri-infarct rim. At two weeks after LAD ligation, rats receiving human angioblasts demonstrated numerous "fingers" of cardiomyocytes of rat origin, as determined by expression of rat MHC class I molecules, extending from the peri-infarct region into the infarct zone. The islands of cardiomyocytes at the peri-infarct rim in animals receiving human angioblasts contained a high frequency of rat myocytes with DNA activity, as determined by dual staining with mAbs reactive against cardiomyocyte-specific troponin I and rat Ki67. In contrast, in animals receiving saline there was a high frequency of cells with fibroblast morphology and reactivity with rat Ki67, but not troponin I, within the infarct zone. The number of cardiomyocytes progressing through cell cycle at the peri-infarct region of rats receiving human angioblasts was 40-fold higher than that at sites distal to the infarct, 20-fold higher than that found in non-infarcted hearts, and 5-fold higher than that at the peri-infarct rim of animals receiving saline.

Neovascularization of Acutely Ischemic Myocardium by Human Angioblasts Prevents Ventricular Remodelling and Causes Sustained Improvement in Cardiac Function. By 15 weeks postinfarction, rats receiving human angioblasts demonstrated markedly smaller scar sizes together with increased mass of viable myocardium within the anterior free wall. Whereas collagen deposition and scar formation extended almost through the entire left ventricular wall thickness in controls, with aneurysmal dilatation and typical EKG abnormalities, the infarct scar extended only to 20-50% of the left ventricular wall thickness in rats receiving CD34+ cells. Moreover, pathological collagen deposition in the non-infarct zone was markedly reduced in rats receiving CD34+ cells. At 15 weeks, the mean proportion of scar/ normal left ventricular myocardium was 13% in rats receiving CD34+ cells compared with 36-45% for each of the other groups studied (saline, CD34-, SVEC).

Remarkably, by two weeks after injection of human angioblasts, and in a parallel time-frame with the observed neovascularization, left ventricular ejection fraction (LVEF) recovered by a mean of 22%. This effect was long-lived, with LVEF recovering by a mean of 34% at the end of follow-up, 15 weeks post injection.

Neither CD34- cells nor SVEC demonstrated similar effects. At 15 weeks post-infarction, mean cardiac index in rats injected with human angioblasts was only reduced by 26% relative to normal rats, whereas for each of the other groups it was reduced by 48-59%.

Together, these results indicate that the neovascularization, reduction in peri-infarct myocyte apoptosis and increase in cardiomyocyte regeneration/ proliferation observed at two weeks prevented myocardial replacement with fibrous tissue and caused sustained improvement in myocardial function.

Potential Use of Angioblasts in Combination with Cardiomyocyte Progenitors for Repair and Regeneration of

Ischemic Myocardium. While increasing capillary density through angioblast-dependent neovascularization is a promising approach for preventing apoptotic death and inducing regeneration of endogenous cardiomyocytes following acute myocardial infarction, the role of angioblast therapy for the treatment of congestive heart failure following chronic ischemia is at present unknown. Nevertheless, it is reasonable to anticipate that cellular therapies for congestive heart failure due to ischemic cardiomyopathy will need to address two interdependent processes: (1) a renewable source of proliferating, functional cardiomyocytes, and (2) development of a network of capillaries and larger size blood vessels for supply of oxygen and nutrients to both the chronically ischemic, endogenous myocardium and to the newlyimplanted cardiomyocytes. To achieve these endpoints it is likely that co-administration of angioblasts and mesenchymal stem cells will be needed in order to develop regenerating cardiomyocytes, vascular structures, and supporting cells such as pericytes and smooth muscle cells. Future studies will need to address the timing, relative concentrations, source and route of delivery of each of these cellular populations in animal models of acute and chronic myocardial ischemia.

In addition to synergistic cellular therapies, it is likely that optimal regimens for the treatment of acute and chronically ischemic hearts will require a combined approach employing additional pharmacological strategies. For example, augmentation in myocardial function might be achieved by combining infusion of human angioblasts and cardiomyocyte progenitors together with beta blockade, ACE inhibition or AT1-receptor blockade to reduce angiotensin II-dependent cardiac fibroblast proliferation and collagen secretion. Understanding the potential of defined lineages of stem cells or undifferentiated progenitors, and their interactions with pharmacological interventions, will lead to better and more focussed clinical trial designs using each cell type independently or in combination, depending on which particular clinical indication is being targeted.

Manufactured products are made from atoms. The properties of those products depend on how those atoms are arranged. If we rearrange the atoms in coal, we get diamonds. If we rearrange the atoms in sand (and add a pinch of impurities) we get computer chips. If we rearrange the atoms in dirt, water and air we get grass. Since we first made stone tools and flint knives we have been arranging atoms in great thundering statistical herds by casting, milling, grinding, chipping and the like. We've gotten better at it: we can make more things at lower cost and greater precision than ever before. But at the molecular scale we're still making great ungainly heaps and untidy piles of atoms. That's changing. In special cases we can already arrange atoms and molecules exactly as we want. Theoretical analyses make it clear we can do a lot more. Eventually, we should be able to arrange and rearrange atoms and molecules much as we might arrange LEGO blocks. In not too many decades we should have a manufacturing technology able to:

- Build products with almost every atom in the right place.
- Do so inexpensively.
- Make most arrangements of atoms consistent with physical law.

Often called *nanotechnology*, *molecular nanotechnology* or *molecular manufacturing*, it will let us make most products lighter, stronger, smarter, cheaper, cleaner and more precise.

The Advantages of Nanotechnology

One of the basic principles of nanotechnology is positional control. At the macroscopic scale, the idea that we can hold parts in our hands and assemble them by properly positioning them with respect to each other goes back to prehistory: we celebrate ourselves as the tool using species. Our wisdom and our knowledge would have done us scant good without an opposable thumb: we'd still be shivering in the bushes, unable to start a fire. At the molecular scale, the idea of holding and positioning molecules is new and almost shocking. However, as long ago as 1959 Richard Feynman, the Nobel prize winning physicist, said that nothing in the laws of physics prevented us from arranging atoms the way we want: "...it is something, in principle, that can be done; but in practice, it has not been done because we are too big." What would it mean if we could inexpensively make things with every atom in the right place?

- For starters, we could continue the revolution in computer hardware right down to molecular gates and wires — something that today's lithographic methods (used to make computer chips) could never hope to do.
- We could inexpensively make very strong and very light materials: shatterproof diamond in precisely the shapes we want, by the ton, and over fifty times lighter than steel of the same strength.
- We could make a Cadillac that weighed fifty kilograms, or a full-sized sofa you could pick up with one hand.
- We could make surgical instruments of such precision and deftness that they could operate on the cells and even molecules from which we are made — something well beyond today's medical technology.

The list goes on — almost any manufactured product could be improved, often by orders of magnitude. Nanotechnology should let us make almost every manufactured product faster, lighter, stronger, smarter, safer and cleaner. We can already see many of the possibilities as these few examples illustrate. New products that solve new problems in new ways are more difficult to foresee, yet their impact is likely to be even greater. Could Edison have foreseen the computer, or Newton the communications satellite?

1. Improved Transportation

- Today, most airplanes are made from metal despite the fact that diamond has a strength-to-weight ratio over 50 times that of aerospace aluminum. Diamond is expensive, we can't make it in the shapes we want, and it shatters. Nanotechnology will let us inexpensively make shatterproof diamond (with a structure that might resemble diamond fibers) in exactly the shapes we want. This would let us make a Boeing 747 whose unloaded weight was 50 times lighter but just as strong.
- Today, travel in space is very expensive and reserved for an elite few. Nanotechnology will dramatically reduce the costs and increase the capabilities of space ships and space flight. The strength-to-weight ratio and the cost of components are absolutely critical to the performance and economy of space ships: with nanotechnology, both of these parameters will be improved... Beyond inexpensively providing remarkably light and strong materials for space ships, nanotechnology will also provide extremely powerful computers with which to guide both those ships and a wide range of other activities in space.

2. Atom Computers

- Today, computer chips are made using lithography — literally, "stone writing." If the computer hardware revolution is to continue at its current pace, in a decade or so we'll have to move beyond lithography to some new post lithographic manufacturing technology. Ultimately, each logic element will be made from just a few atoms.
- Designs for computer gates with less than 1,000 atoms have already been proposed — but each atom in such a small device has to be in exactly the right place. To economically build and interconnect trillions upon trillions of such small and precise devices in a complex three dimensional pattern we'll need a manufacturing technology well beyond today's lithography: we'll need nanotechnology.
- With it, we should be able to build mass storage devices that can store more than a hundred billion billion bytes in a volume the size of a sugar cube; RAM that can store a mere billion billion bytes in such a volume; and massively parallel computers of the same size that can deliver a billion billion instructions per second.

3. Military Applications

- Today, "smart" weapons are fairly big — we have the "smart bomb" but not the "smart bullet." In the future, even weapons as small as a single bullet could pack more computer power than the largest supercomputer in existence today, allowing them to perform real time image analysis of their surroundings and communicate with weapons tracking systems to acquire and navigate to targets with greater precision and control.

- We'll also be able to build weapons both inexpensively and much more rapidly, at the same time taking full advantage of the remarkable materials properties of diamond. Rapid and inexpensive manufacture of great quantities of stronger more precise weapons guided by massively increased computational power will alter the way we fight wars. Changes of this magnitude could destabilize existing power structures in unpredictable ways. Military applications of nanotechnology raise a number of concerns that prudence suggests we begin to investigate before, rather than after, we develop this new technology.

4. Solar Energy

- Nanotechnology will cut costs both of the solar cells and the equipment needed to deploy them, making solar power economical. In this application we need not make new or technically superior solar cells: making inexpensively what we already know how to make expensively would move solar power into the mainstream.

5. Medical Uses

- It is not modern medicine that does the healing, but the cells themselves: we are but onlookers. If we had surgical tools that were molecular both in their size and precision, we could develop a medical technology that for the first time would let us directly heal the injuries at the molecular and cellular level that are the root causes of disease and ill health. With the precision of drugs combined with the intelligent guidance of the surgeon's scalpel, we can expect a quantum leap in our medical capabilities.

The single most frequently asked question about nanotechnology is: How long? How long before it will let us make molecular computers? How long before inexpensive solar cells let us use clean solar power instead of oil, coal, and nuclear fuel? How long before we can explore space at a reasonable cost? The scientifically correct answer is: *I don't know*. From relays to vacuum tubes to transistors to integrated circuits to Very Large Scale Integrated circuits (VLSI) we have seen steady declines in the size and cost of logic elements and steady increases in their performance.

- Extrapolation of these trends suggests we will have to develop molecular manufacturing in the 2010 to 2020 time frame if we are to keep the computer hardware revolution on schedule.
- Of course, extrapolating past trends is a philosophically debatable method of technology forecasting. While no fundamental law of nature prevents us from developing nanotechnology on this schedule (or even faster), there is equally no law that says this schedule will not slip.
- Much worse, though, is that such trends imply that there is some ordained schedule — that nanotechnology will appear regardless of what we do or don't do. Nothing could be further from the truth. How long it takes to develop this technology depends very much on what we do. If we pursue it systematically, it will happen sooner. If we ignore it, or simply hope that someone will stumble over it, it will take much longer. And by using theoretical, computational and experimental approaches together, we can reach the goal more quickly and reliably than by using any single approach alone.

While some advances are made through serendipitous accidents or a flash of insight, others require more work. It seems unlikely that a scientist would forget to turn off the

Bunsen burner in his lab one afternoon and return to find he'd accidentally made a Space Shuttle. Like the first human landing on the moon, the Manhattan project, or the development of the modern computer, the development of molecular manufacturing will require the coordinated efforts of many people for many years. How long will it take? A lot depends on when we start.

4

Stem Cell Research in USA: A Law and Policy Perspective

Current Cell Research Law and Policy Regarding Stress

Any overview of the state of human stem cell research under the current federal funding policy must begin with a thorough understanding of that policy. This is not as simple as it may sound. From the moment of its first announcement, on August 9, 2001, the policy has been misunderstood (and at times misrepresented) by some among both its detractors and its advocates. Its moral foundation, its political context, its practical implications, and the most basic facts regarding the policy's implementation have all been subjects of heated dispute and profound confusion. Whether one agrees with the policy or not, it is important to understand it as it was propounded, accurately and in its own terms, in the light also of the historical and political contexts in which it was put forward.

This chapter attempts to place the policy in its proper context; to articulate its moral, legal, and political underpinnings (as put forward by its authors and advocates); to offer an overview of its implementation thus far; and to begin to describe its ramifications for researchers and for medicine.By articulating the policy in its own terms, we intend neither to endorse it nor to find fault with it. Some

Members of the Council oppose the current policy, some Members support it. Yet the descriptive account that we offer here aspires to be seen as accurate and fair, regardless of where one personally stands on the issue. Nearly all Members of this Council recognize, as we said in our report *Human Cloning and Human Dignity*, that "all parties to this debate have something vital to defend, something vital not only to themselves but also to their opponents in the debate, and indeed to all human beings. No human being and no human society can afford to be callous to the needs of suffering humanity, cavalier regarding the treatment of nascent human life, or indifferent to the social effects of adopting in these matters one course of action rather than another." (*Human Cloning and Human Dignity,* p 121.) Thus, whatever we think of the current funding policy, we recognize that this is a genuine ethical dilemma and that reasonable people of good will may come to different conclusions about where the best ethical or policy position lies. We therefore also believe that not only results but also reasons matter, and that it behooves us to understand the principled or prudential reasons for the current policy (as well as for any alternative policy that might be offered to replace it). Indeed, in the next chapter we present an overview of arguments on all sides of the question. Here we mean only to clarify, as far as we are able, the original meaning and purpose of the policy, so as to be better able to monitor its impact.

A Brief History of the Embryo Research Funding Debate

The federal government makes significant public resources available to biomedical researchers each year—over $20 billion in fiscal year 2003 alone—in the form of research grants offered largely through the National Institutes of Health (NIH). This level of public expenditure reflects the great esteem in which Americans hold the biomedical enterprise and the value we place on the development of treatments and cures for those who are suffering. But such support is not offered indiscriminately. Researchers who accept federal funds must abide by ethically based rules and regulations governing, among other things, the use of human subjects in research.

And some policymakers and citizens have always insisted that taxpayer dollars not be put toward specific sorts of research that violate the moral convictions and sensibilities of some portion of the American public. This has meant that controversies surrounding the morality of some forms of scientific research have at times given rise to disputes over federal funding policy. Among the most prominent examples has been the three-decade-long public and political debate about whether taxpayer funds should be used to support research that involves creating or destroying human embryos or making use of destroyed embryos and fetuses—practices that touch directly on the much-disputed questions of the moral status and proper treatment of nascent human life.

In the immediate aftermath of the Supreme Court's 1973 *Roe v. Wade* decision legalizing abortion nationwide, some Americans, including some Members of Congress, became concerned about the potential use of aborted fetuses (or embryos) in scientific research. In response to these concerns, the Department of Health, Education and Welfare (DHEW, the precursor to today's Department of Health and Human Services) initiated a moratorium on any potential DHEW sponsorship or funding of research using human fetuses or living embryos. In 1974, Congress codified the policy in law, initiating what it termed a temporary moratorium on federal funding for clinical research using "a living human fetus, before or after the induced abortion of such fetus, unless such research is done for the purpose of assuring the survival of such fetus." Concurrently with that moratorium (and also addressing concerns not directly related to embryo and fetal research), Congress established a National Commission for the Protection of Human Subjects of Biomedical and Behavioral Research. Among its other tasks, Congress explicitly assigned the Commission responsibility for offering guidelines for human fetal and embryo research, so that standards for funding might be established and the blanket moratorium might be lifted. The statutory moratorium was lifted once the Commission issued its report in 1975.

In that report, the Commission called for the establishment of a national Ethics Advisory Board within DHEW to propose standards and research protocols for potential federal funding of research using human embryos and to consider particular applications for funding. In doing so, the Commission looked ahead to the possible uses of in vitro embryos, since the first successful in vitro fertilization (IVF) of human egg by human sperm had been accomplished in 1969. In its discussion of "fetal" research, the commission defined the fetus as the product of conception from the time of implantation onward, which therefore included what we generally think of (and define in this report) as embryos in utero. Its separate consideration of embryo research was therefore directed at in vitro embryos.The Department adopted the recommendation in 1975, established an Ethics Advisory Board, and put in place regulations requiring that the Board provide advice about the ethical acceptability of IVF research proposals. The Board first took up the issue of research on in vitro embryos in full in the late 1970s and issued its report in 1979. By that time, human IVF techniques had been developed (first in Britain) to the point of producing a live-born child (born in 1978). These techniques, and their implications for human embryo research, raised unique prospects and concerns that were distinct from some of those involved in human fetal research. As a consequence, starting in the late 1970s, funding of embryo research and funding of fetal research came to be treated as mostly distinct and separate issues. The Ethics Advisory Board concluded that research involving embryos and IVF techniques was "ethically defensible but still legitimately controverted." Provided that research did not take place on embryos beyond fourteen days of development and that all gamete donors were married couples, the Board argued, such work was "acceptable from an ethical standpoint," but the Board decided that it "should not advise the Department on the level of Federal support, if any," such work should receive. This left the decision in the hands of the DHEW, which decided at that stage not to offer funding for human embryo studies. The Ethics Advisory Board's charter expired in 1980, and no renewal or replacement was put forward, creating a peculiar situation in which the regulations requiring the Ethics Advisory Board to review proposals for funding remained in effect, but the Board no longer existed to consider such requests. Funding was therefore rendered impossible in practice. Because the Ethics Advisory Board was never replaced, a de facto ban on funding remained in place through the 1980s.

In 1993, Congress enacted the NIH Revitalization Act, a provision of which rescinded the requirement that research protocols be approved by the non-existent Ethics Advisory Board. This change opened the way in principle to the possibility of NIH funding for human embryo research using IVF embryos. The following year, the NIH convened a Human Embryo Research Panel to consider the issues surrounding such research and to propose guidelines for potential funding applications. The panel recommended that some areas of human embryo research be deemed eligible for federal funding within a framework of recognized ethical safeguards. It further concluded that the creation of human embryos with the explicit intention of using them only for research purposes should be supported under some circumstances. President Clinton overruled the panel on the latter point, ordering that embryo creation for research not be funded, but he accepted the panel's other recommendations and permitted the NIH to consider applications for funding of research using embryos left over from IVF procedures. Congress, however, did not endorse this course of action.

In 1995, before any funding proposal had ever been approved by the NIH, Congress attached language to the 1996 Departments of Labor, Health and Human Services, and Education, and Related Agencies Appropriations Act (the budget bill that funds DHHS and the NIH) prohibiting the use of any federal funds for research that destroys or seriously endangers human embryos, or creates them for research purposes.

This provision, known as the "Dickey Amendment" (after its original author, former Representative Jay Dickey of Arkansas), has been attached to the Health and Human Services appropriations bill each year since 1996. Everything about the subsequent debate over federal funding of embryonic stem cell research must be understood in the context of this legal restriction. The provision reads as follows: None of the funds made available in this Act may be used for— (1) the creation of a human embryo or embryos for research purposes; *or* (2) research in which a human embryo or embryos are destroyed, discarded, or knowingly subjected to risk of injury or death greater than that allowed for research on fetuses in utero under 45 CFR 46.204 and 46.207, and subsection 498(b) of the Public Health Service Act (42 U.S.C. 289g(b)).(b) For purposes of this section, the term 'human embryo or embryos' includes any organism, not protected as a human subject under 45 CFR 46 as of the date of the enactment of the governing appropriations act, that is derived by fertilization, parthenogenesis, cloning, or any other means from one or more human gametes or human diploid cells. This law effectively prohibits the use of federal funds to support any research that destroys human embryos or puts them at serious risk of destruction. It does not, however, prohibit the conduct of such research using private funding.

Thus, it addresses itself not to what may or may not be lawfully done, but only to what may or may not be supported by taxpayer dollars. At the federal level, research that involves the destruction of embryos is neither prohibited nor supported and encouraged. The Dickey Amendment was originally enacted before the isolation of human embryonic stem cells, first reported in 1998 by researchers at the University of Wisconsin, whose work was supported only by private funds (largely from the Geron Corporation and the University of Wisconsin Alumni Research Foundation). The discovery of these cells and their unique and potentially quite promising properties aroused great excitement both within and beyond the scientific community. It led some people to question the policy of withholding federal funds from human embryo research. Most Members of Congress, however, did not change their position, and the Dickey Amendment has been reenacted every year since. For many of its supporters, the amendment expresses their ethical conviction that nascent human life ought to be protected against exploitation and destruction for scientific research, however promising that research might be, and that at the very least such destruction should not be supported or encouraged by taxpayer dollars.

On its face, the Dickey Amendment would seem to close the question of federal funding of human embryonic stem cell research, since obtaining stem cells for such research relies upon the destruction of human embryos. But in 1999, the General Counsel of the Department of Health and Human Services argued that the wording of the law might permit an interpretation under which human embryonic stem cell research could be funded. If embryos were first destroyed by researchers supported by private funding, then subsequent research employing the derived embryonic stem cells, now propagated in tissue culture, might be considered eligible for federal funding. Although such research would presuppose and follow the prior destruction of human embryos, it would not itself involve that destruction. Thus, the Department's lawyers suggested, the legal requirement not to fund research "in which" embryos were destroyed would still technically be obeyed.9 This has generally been taken to be a legally valid interpretation of the specific language of the statute, and indeed the subsequent policies of both the Clinton and Bush administrations have relied upon it in different ways. But some critics of the 1999 legal opinion argued that, though it might stay within the letter of the law, the proposed approach would contradict both the spirit of the law and the principle that underlies it. It would use public funds to encourage and reward the destruction of human embryos by promising funding for research that immediately follows and results from that destruction—thereby offering a financial incentive to engage in such destruction in the future. By so doing, these critics argued, it would at least implicitly state, in the name of the American people, that research that destroys human embryos ought to be encouraged in the cause of medical advance. Supporters of the Clinton administration's proposed approach, however, argued that promoting such research— especially given its therapeutic potential—was indeed an appropriate government function, and that the policy proposed by DHHS was neither illegal nor improper, given the text of the statute and provided that the routine standards of research ethics (including informed consent and a prohibition on financial inducements) were met.

The Clinton administration adopted this course of action and drew up specific guidelines to enact it. But the guidelines, completed just before the end of the Clinton administration, never had a chance to be put into practice, and no funding was ever provided. Upon entering office in 2001, the Bush administration decided to take another look at the options regarding human embryonic stem cell research policy and therefore put the new regulations on hold, pending review.

In conducting its review, the Bush administration stated that it sought a way to allow some potentially valuable research to proceed while upholding the spirit (and not just the letter) of the Dickey Amendment, a spirit that the President himself has advocated. The expressed hope was that the government, while continuing to withhold taxpayer support or encouragement for the destruction of human embryos, might find a way to draw some moral good from stem cell lines that had already been produced through such destruction—given that this deed, even if immoral, could not now be undone. This is the ethical-legal logic of the present stem cell funding policy: it seeks those benefits of embryonic stem cell research that might be attainable without encouraging or contributing to any future destruction of human embryos.

The Present Policy

The current policy on federal government funding of human embryonic stem cell research, then, must be understood in terms of the constraints of the Dickey Amendment and in terms of the logic of the moral and political aims that underlie that amendment. At the time of the policy's announcement, a number of embryonic stem cell lines had already been derived and were in various stages of growth and characterization. The embryos from which they were derived had therefore already been destroyed and could no longer develop further. As President Bush put it, "the life and death decision had already been made." The administration's policy made it possible to use taxpayer funding for research conducted on those preexisting lines, but it refused in advance to support research on any lines created after the date of the announcement. In addition, to be eligible for funding, those preexisting lines would have had to have been derived from excess embryos created solely for reproductive purposes, made available with the informed consent of the donors, and without any financial inducements to the donors—standard research-ethics conditions that had been attached to the previous administration's short-lived funding guidelines, as well as to earlier attempts to formulate rules for federal funding of human embryo research. The policy denies federal funding not only for research conducted on stem cell lines derived from embryos destroyed after August 9, 2001 (or that fail to meet the above criteria), but also (as the proposed Clinton-era policy would have) for the creation of any human embryos for research purposes and for the cloning of human embryos for any purpose. The moral, legal, and political grounds of this policy have been hotly contested from the moment of its announcement.

Debates have continued regarding its aims, its character, its implementation, and its underlying principles, as well as the significance of federal funding in this area of research. For example, many scientists, physicians, and patient advocacy groups contend that the policy is too restrictive and thwarts the growth of a crucial area of research. On the other side, some opponents of embryo research believe the policy is too liberal and legitimates and rewards (after the fact) the destruction of nascent human life. Some ethicists argue that there is a moral imperative to remove all restrictions upon potentially life-saving research; other ethicists argue that there is a moral imperative to protect the lives of human beings in their earliest and most vulnerable stages. These and similar arguments are reviewed in the next chapter. But before one can enter into these debates, it is essential first to understand the relevant elements of the policy itself as clearly and distinctly as possible.

Moral Foundation of the Policy

In articulating its proposed funding policy in 1999 and 2000, the Clinton administration expressed a firm determination that funded research could use only those human embryos that had been left over from IVF procedures aimed at reproduction and that had been donated in accordance with the standards of informed consent and in circumstances free of financial inducements. Provided that these crucial conditions were met, the administration argued that the potential benefits of stem cell research were so great that publicly funded research should go forward. In August of 2000, reflecting on the guidelines put forward by his administration, President Clinton remarked, Human embryo research [as approved for funding by the NIH guidelines] deals only with those embryos that were, in effect, collected for in-vitro fertilization that never will be used for that. So I think that the protections are there; the most rigorous scientific standards have been met. But if you just—just in the last couple of weeks we've had story after story after story of the potential of stem cell research to deal with these health challenges. And I think we cannot walk away from the potential to save lives and improve lives, to help people literally to get up and walk, to do all kinds of things we could never have imagined, as long as we meet rigorous ethical standards. Given the promise of embryonic stem cell research, the existence of many embryos frozen in IVF clinics and unlikely ever to be transferred and brought to term, and the willingness of some IVF patients to donate such embryos for research, the Clinton administration reasoned that research using cell lines derived from these embryos could ethically be supported by federal funds. That position implies, of course, that the destruction of embryos is not inherently or necessarily unethical, or so disconcerting as to be denied any federal support. The Clinton-era NIH Embryo Research Panel put succinctly one form of this view in stating that "the preimplantation human embryo warrants serious moral consideration as a developing form of human life, but it does not have the same moral status as infants and children." If there is sufficient promise or reason to support research, the claim of a human embryo to "serious moral consideration" (or, as others, including some of us, have put it, to "special respect") could be outweighed by other moral aims or principles.

This (at least implicit) understanding of the moral status of human embryos might be seen to have put the Clinton administration at odds with the principle animating the operative law on this subject (the Dickey Amendment). But given its responsibility to carry out the laws as they are enacted, the administration sought a way to advance research within the limitations set by the statute. Its approach to the funding of embryonic stem cell research, therefore, seems to have sought an answer to this question: *How can embryonic stem cell research, conducted in accordance with standards of informed consent and free donation, be maximally aided within the limits of the law?* The NIH guidelines published in 2000 represent the answer the Clinton administration found: funding research on present and future embryonic stem cell lines, so long as the embryo destruction itself is done with private funds.

The Bush administration appears to have been motivated by a somewhat different question, arising from what seems to be a different view of the morality of research that destroys human embryos. President Bush put the matter this way, in discussing his newly announced policy in August of 2001: Stem cell research is still at an early, uncertain stage, but the hope it offers is amazing: infinitely adaptable human cells to replace damaged or defective tissue and treat a wide variety of diseases. Yet the ethics of medicine are not infinitely adaptable. There is at least one bright line: We do not end some lives for the medical benefit of others. For me, this is a matter of conviction: a belief that life, including early life, is biologically human, genetically distinct and valuable. Using similar language, but speaking even more unambiguously, President Bush reiterated his ethical view of the destruction of human embryos for medical research in a speech on human cloning legislation, saying, "I believe all human cloning is wrong, and both forms of cloning ought to be banned, for the following reasons. First, anything other than a total ban on human cloning would be unethical. Research cloning would contradict the most fundamental principle of medical ethics, that no human life should be exploited or extinguished for the benefit of another. Yet a law permitting research cloning, while forbidding the birth of a cloned child, would require the destruction of nascent human life." ("Remarks by the President on Human Cloning Legislation," as made available by the White House Press Office, April 10, 2002.) While expressing a desire to advance medical research, this argument describes a line that such research should not cross, and therefore past which funding should not be offered. That line, in this context, is the destruction of a human embryo for research purposes. The Bush administration thus appears to share the view that underlies both the word and spirit of the Dickey Amendment. In its approach to the stem cell issue it has sought to answer a question that differs, subtly but significantly, from that formulated by the previous administration: *How can embryonic stem cell research, conducted in accordance with basic research ethics, be maximally aided within the bounds of the principle that nascent human life should not be destroyed for research?* In seeking to answer that question, the Bush administration (like the Clinton administration) had to take account of the existing situation and—as always in such instances—to mix prudential demands and opportunities with an effort at principled judgment. Given the existence of some human embryonic stem cell lines, derived from human embryos that had already been destroyed, the administration determined that it might not simply have to choose between funding research that relies on the ongoing destruction of embryos (and therefore tacitly supporting and encouraging such destruction by paying for the work that immediately follows it) and funding no human embryonic stem cell research at all.

The decision regarding the funding of research on already-derived human embryonic stem cells came down to this question: *Can the government support some human embryonic stem cell research without encouraging future embryo destruction?* The present funding policy is therefore *not* an attempt to answer the question of how the government might best advance embryonic stem cell research while conforming to the law on the subject. Rather, it is an attempt to answer the question of how the government might avoid encouraging the (presumptively) unethical act of embryo destruction and still advance the worthy cause of medical research. Whether or not one agrees with the premises defining the question, and whether or not one accepts the logic of the answer, any assessment of the policy must recognize this starting point.

From the very beginning, the policy has been described—even by many of its supporters and defenders—as occupying a kind of middle-ground position in the debate over the morality of embryo research. It has been termed a "Solomonic compromise." But while it may be a prudential compromise on the question of funding, it has been argued that the policy— as articulated by its authors—does not seem to be intended as a compromise on the question of the moral status of human embryos or the moral standing of the act of embryo destruction.

In this sense, it appears to be not a political "splitting of the difference" but an effort at a principled solution. To some extent, the effort reflects a traditional approach in moral philosophy to an ancient and vexing question: Can one benefit from the results of (what one believes to be) a past immoral act without becoming complicit in that act? The moralists' approach suggests that one may make use of such benefits if (and only if) three crucial conditions are met: (1) *Non-cooperation*: one does not cooperate or actively involve oneself in the commission of the act; (2) *Non-abetting*: one does nothing to abet or encourage the repetition of the act, for instance by providing incentives or rewards to those who would perform it in the future; and (3) *Reaffirmation of the principle*: in accepting the benefit, one re-enunciates and reaffirms the principle violated by the original deed in question.

As a plan for redeeming some good from embryo destruction that has already taken place, while not encouraging embryo destruction in the future, the administration's policy appears at least to seek to address each of these three conditions: (1) No federal funds have been or, by this policy, would be used in the destruction of human embryos for research. (2) By restricting research funding exclusively to embryonic stem cell lines derived before the policy went into effect, the policy deliberately refuses to offer present or future financial or other incentives to anyone who might subsequently destroy additional embryos for research; this is the moral logic behind a central feature of the policy, the cut-off date for funding eligibility (though some argue that by failing to call for an end to privately funded research the policy does not altogether avoid complicity). And (3) the President, in his speech of August 9, 2001, and since (as in the passage quoted above and elsewhere), has reaffirmed the moral principle that underlies his policy and the law on the subject: that nascent human life should not be destroyed for research, even if good might come of it. The policy as a whole draws attention to that principle by drawing a sharp line beyond which funding will not be made available.

Of course, since these terms from the parlance of moral philosophy were not those explicitly employed by the policy's authors, they can go only so far in helping us to understand the policy's foundation. As in any public policy decision, prudence is here mixed with principle, in the hope that the two might reinforce (rather than undermine) each other, and a variety of moral aims are brought together. The desire to afford some aid to a potentially promising field of research moderates what might otherwise have been an at least symbolically stauncher stance against embryo destruction: no public funding whatsoever, even for work on stem cell lines obtained from embryos destroyed in the past. Moreover, the desire to show regard for established principles and standards of ethical research leads to an insistence that, to be approved, stem cell lines must have been drawn from embryos produced for reproduction and obtained with consent and without financial inducements. In these ways, the policy gives some due to competing moral and prudential demands. But the policy's central feature—the announcement date separating eligible from ineligible stem cell lines—holds firm to the principle that *public funds* should not be used to encourage or support the destruction of embryos *in the future*.

It is perhaps worth pointing out that one's attitude regarding the best federal funding policy is not simply determined by one's view regarding the moral standing of human embryos, and that even persons who hold the same view of the moral standing of human embryos may not all agree on the best policy. For example, support for the current policy does not *necessarily* require a belief that human embryos are persons with full moral standing; and conversely, those who believe that human embryos are persons do not necessarily support the policy. One might believe, for instance, that an embryo is a mystery, not clearly "one of us" but unambiguously a life-in-process, and thus conclude that we should err on the side of restraint (non-destruction) when moral certainty is impossible. Or, one might believe that embryos are not simply persons but are nonetheless either worthy of protection from harm or at least worthy of more respect than ordinary human tissues or animals, and that it would be wrong to begin a massive public project of embryo research that offends the deeply-held beliefs of many citizens. Meanwhile, an individual who believes that human embryos have the same moral standing as children or adults may be deeply unsatisfied with the present policy, since merely denying federal encouragement for future embryo destruction while taking no action to prevent privately-funded stem cell research that destroys embryos may be an insufficient response to the ongoing destruction of nascent human life.

For some of its supporters, the policy goes as far as it seems possible to go within the bounds of the spirit and aims of the law—that the government should not encourage or support the destruction of nascent human life for research. Yet at the same time, it goes farther than the federal government has gone before in the direction of actually funding research involving human embryos, since no public funds had ever before been spent on such research. To go further—say, by funding research on the currently ineligible lines derived after August 9, 2001—would not extend the logic of the policy or of the law, but rather contradict them both: it would be a difference not of degree but of principle.

By implying that research using embryos destroyed in the future might one day be supported with public funding, such a policy shift would at least implicitly encourage the very act (embryo destruction) that the current policy aims not to encourage. Of course, such a change might well be in order, but the case for it must address itself to the moral argument and its principles, and not only to the state of research and its progress or promise.

Rather than focus on this principled aspect of the policy, the public debate has tended to concentrate on the precise balance of benefits and harms resulting from the combination of the administration's policy and the state of the relevant science. It has focused on whether there are "enough" cell lines or on whether the science is advancing as quickly as it could. And it has proceeded as though the administration's aim was simply to maximize progress in embryonic stem cell research without transgressing the limits of the letter of the law.

Had the decision been based on that aim alone, then claims or evidence of slowed progress alone might, in themselves, constitute an effective argument against it on its own terms (on the ground that the law technically permits federal funding of research on cells derived from embryos whose destruction was underwritten by private funding).

But if one accepts the premise that the decision was grounded also in a discernible (albeit highly controversial) moral aim, one cannot show that the policy is wrong merely by pointing to the potential benefits of stem cell research or the potential harm to science caused by restrictions in federal funding. The present policy aims to support stem cell research while insisting that federal funds not be used to support or encourage the future destruction of human embryos. To argue with that policy on its own terms, therefore, one would need to argue with its view of the significance of that aim. Concretely, this means arguing with its ethical position regarding the destruction of nascent human life and with its ethical-political position regarding the significance of government funding of a contested activity.

This latter point—regarding the meaning of government funding—is much neglected in the current debates and deserves further clarification. That will require delving into the important distinction between government permission (that is, an absence of prohibitions) of an activity and government support for an activity. This ethical-political distinction lies at the heart of the stem cell debate.

The Significance of Federal Funding

The national debate over human embryonic stem cell research often raises the most fundamental questions about the moral status of human embryos and the legitimacy of research that destroys such embryos. Yet, looking over this debate, it is easy to forget that the question at issue is not whether research using embryos should be allowed, but rather whether it should be financed with the federal taxpayer's dollars.

The difference between *prohibiting* embryo research and *refraining from funding* it has often been blurred by both sides to the debate. Ignored in the battles over embryo research itself, the ethical-political question regarding funding is rarely taken up in full.

That question arises because modern governments do more than legislate and enforce prohibitions and limits. In the age of the welfare state, the government, besides being an enforcer of laws and a keeper of order, is also a major provider of resources. Political questions today, therefore, reach beyond what ought and ought not be allowed. They include questions of what ought and ought not be encouraged, supported, and made possible by taxpayer funding. The decision to fund an activity is more than an offer of resources. It is also a declaration of official national support and endorsement, a positive assertion that the activity in question is deemed by the nation as a whole, through its government, to be good and worthy.

When something is done with public funding, it is done, so to speak, in the name of the country, with its blessing and encouragement. To offer such encouragement and support is therefore no small matter. The federal government is not required to provide such material support, even for activities protected by the Constitution, let alone for those permitted but not guaranteed. The affording of most federal funding is entirely optional, and the choice to make such an offer is therefore laden with moral and political meaning, well beyond its material importance.

In the age of government funding, the political system is sometimes called upon to decide not only the lowest standards of conduct, but also the highest standards of legitimacy and importance. When the nation decides an activity is worth its public money, it declares that the activity is valued, desired, and favored.

The United States has long held the scientific enterprise in such high regard. Since the middle of the twentieth century, the federal government, with the strong support of the American people, has funded scientific research to the tune of many hundreds of billions of dollars. The American taxpayer is by far the greatest benefactor of science in the world, and the American public greatly values the contributions of science to human knowledge, human health, and human happiness.

And we Americans have overwhelmingly been boosters of medical science and medical progress, deeming them worthy of support for moral as well as material reasons.

But this enthusiasm for medical science is not without its limits. As already noted, we attach restrictions to federally funded research, for instance to protect human subjects. In fact at times we even use funding to *place* restrictions on research that might otherwise not be constrained. Indeed, federal funding sometimes serves as a means by which *private* research can be subjected to critical standards, since institutions that receive federal funds are often inclined (and given strong administrative incentives) to abide by the prescribed ethical standards throughout all of their activities, not only those directly receiving public dollars. Some supporters of funding therefore argue that extending public money to research is the most effective means of making certain that nearly all researchers, public and private, adhere to basic standards of ethics and safety. Public funding also requires researchers to make their work available to the public and for critical review by their peers, and it may encourage some degree of responsibility not necessarily encouraged by commercial endeavors. In addition to conditions attached to government funding of research, law sometimes erects specific limits on certain practices that might be medically beneficial. For example, we put limits on some practices that might offer life-saving benefits, such as the buying and selling of organs for transplantation, currently prohibited under the National Organ Transplant Act. Also, as in the present case, many Americans and their congressional representatives have moral reasons for opposing certain lines of research or clinical practice, for example those that involve the exploitation and destruction of human fetuses and embryos.

The two sides of the embryo research debate tend to differ sharply on the fundamental moral significance of the activity in question. One side believes that what is involved is morally abhorrent in the extreme, while the other believes embryo research is noble or even morally obligatory and worthy of praise and support. It would be very difficult for the government to find a middle ground between these two positions, since the two sides differ not only on what should or should not be done, but also on the moral premises from which the activity should be approached.

To this point, the federal government has pursued a policy whereby it does not explicitly prohibit embryo research but also does not officially condone it, encourage it, or support it with public funds. This approach, again, combines prudential demands with moral concerns. It has allowed the political system to avoid banning embryo research against the wishes of those who believe it serves an important purpose, while not compelling those citizens who oppose it to fund it with their tax money.

This approach is also based, at least in part, on the conviction that debates over the federal budget are not the place to take up the anguished question of the moral status of human embryos. But the position is not only a compromise between those who would have the government bless and those who would have the government curse this activity. It is also a statement of a certain principle: namely, that public sanction makes a serious difference and ought not to be conferred lightly. While embryo destruction may be something that some Americans support and engage in, it is not something that America *as a nation* has officially supported or engaged in.* Of course, if the funding issue were merely a proxy for the larger dispute over the moral status of human embryos, then the present arrangement might appeal only to those who would protect human embryos, and it would succeed only as long as they were able to enact it. The argument might end there, with a vote-count on the question of the moral status or standing of human embryos. But some proponents of the present law suggest that the particulars and contours of the embryo research debate offer an additional rationale for that arrangement. Here again, it is important to remember that the issue in question is public funding, not permissibility. Opponents of embryo research have in most cases acquiesced (likely owing to various prudential and moral factors) in narrowing the debate at the federal level to the question of funding. They do not argue for a wholesale prohibition of embryo research by national legislation, even though many of them see such work as an abomination and even a form of homicide. In return, proponents of the Dickey Amendment argue that it would be appropriate for supporters of research to agree to do without federal funding in this particular field.

On the other hand, it might reasonably be argued that part of living under majority rule is living with the consequences of sometimes being in the minority. Were the Congress to overturn the current policy of withholding public funds from the destruction of embryos, opponents of funding for embryo research would not be alone in being compelled to pay for activities they abhor. We all see our government do things, in our name, with which we disagree. Some of these might even involve life and death questions of principle, for instance in waging wars that some citizens deeply oppose.

The existence of strong moral opposition to some policy is not in itself a decisive argument against proceeding with that policy.

These concerns give the question of funding its own crucial ethical significance, even apart from the more fundamental question of the legitimacy and propriety of the act being funded. This matter of funding broadly understood, together with the moral and prudential aims apparently motivating the administration's policy, as well as the legal context created by the Dickey Amendment, are the essential prerequisites for thinking about the underlying logic of the current policy. The combination of these elements gives form not only to the specific rules set forth in the administration's funding policy, but also to the implementation of that policy, to which subject we now turn.

Implementation of the Present Policy

The complex and critical task of implementing the funding policy falls largely to the National Institutes of Health, which administers most federal funding of biomedical research. As noted, the administration's policy attempts to advance stem cell research within the bounds already laid out regarding further destruction of human embryos. Thus, while the funding criteria of the policy set the bounds, the NIH, in its ongoing work, is expected to advance the goal of maximally effective funding and support within those bounds.

To this end, the NIH has worked to "jump-start" this field of research through a series of coordinated activities. To plan and oversee these activities, the NIH has established a Stem Cell Task Force charged with determining the best uses for public funds in the field and with putting in place the resources required to make effective use of those funds. The most basic material resources in question are the human embryonic stem cell lines themselves. In August 2001, President Bush announced that "more than sixty genetically diverse stem cell lines" (or stem cell preparations) already existed, and so would be eligible for funding under his policy.

The NIH now believes the actual number to be somewhat higher, so that seventy-eight lines (or preparations) are known to be eligible for funding. These numbers took almost everyone by surprise. Prior to the President's announcement, the best estimates of the number of human embryonic stem cell lines then existing worldwide ranged between 10 and 20. But eligibility is not the same thing as availability, as we will discuss. The lines are held by universities, companies, and other entities throughout the world. According to the National Institutes of Health's latest report (September 2003), the following organizations have developed stem cell derivations eligible for federal funding (that is, derived prior to August 9, 2001, under the approved conditions):

Although all of these lines (or preparations) are deemed *eligible* for funding according to the criteria of the administration's policy, not all are presently *available* for use by researchers (nor is it clear that *all* of them will ever be available for widespread use). Indeed, a point critical to understanding the current situation is that as of the autumn of 2003 only *twelve* lines are available for use,22 while most of the other lines are not yet adequately characterized or developed (some exist only as frozen stocks) and so have at least not yet become available. The process of establishing a human embryonic stem cell line, turning the originally extracted cells into stable cultured populations suitable for distribution to researchers, involves an often lengthy process of growth, characterization, quality control and assurance, development, and distribution. In addition, the process of making lines available to federally funded researchers involves negotiating a contractual agreement (a "materials transfer agreement") with the companies or institutions owning the cell lines, establishing guidelines for payment, intellectual property rights over resulting techniques or treatments, and other essential legal assurances between the provider and the recipient.

The entire process—scientific and legal—has tended to take at least a year for each cell line. Thus, determining which of the 78 eligible lines are in sufficiently good condition, characterizing and developing those lines, and establishing the arrangements necessary to make them available has been a demanding task. By September of 2003, slightly over two years after the enactment of the funding policy, twelve of the eligible lines had become available to federally funded researchers. The NIH has made available "infrastructure award" funds (totaling just over $6 million to date) to a number of the institutions that possess eligible cell-lines, to enable them to more quickly and effectively develop more lines to distribution quality.

As a result, while the number of available lines (only one in the summer of 2002 but risen to twelve in the autumn of 2003) is expected to continue to grow with time, it is unclear how many of the 78 lines will finally prove accessible and useful. According to the NIH, as of the autumn of 2003, the owners of the available lines have distributed over 300 shipments of lines to researchers. No information is presently available on the number of individual researchers or institutions that have received lines. Successful implementation of the current funding policy depends not only on the availability of eligible lines, but also on adequate allocation of financial resources to develop and make use of those lines and to advance the field in general. The funding policy, though it limits the targets of funding to the eligible lines, does not directly delimit or restrict the *amount of money* and other resources that the NIH may invest in human embryonic stem cell research. The amount invested, a decision left to NIH and the Congressional appropriations process, is largely a function of the number of qualified applicants for funding and of the NIH's own priorities and funding decisions. Of course, if more lines were eligible for funding, it is quite possible that more funding would be allocated, but the *amount* that *can* be allocated to work on existing lines is not limited by the funding criteria. In fiscal year 2002, the NIH devoted approximately $10.7 million to human embryonic stem cell research. Based upon a September 2003 estimate, it will have spent approximately $17 million in fiscal year 2003. It is expected that further increases will follow as the field and the number of grant applications grow.

As of November 2003, NIH funds have been allocated to support the following new and continuing awards for human embryonic stem cell research: nine infrastructure awards to assist stem cell providers to expand, test, and perform quality assurance, and improve distribution of cell lines that comply with the administration's funding criteria (aimed at making more of the eligible lines available); 28 investigator-initiated awards for specific projects; 88 administrative supplements (awarded to scientists already receiving funds for work on other sorts of stem cells, either non-embryonic or non-human, to enable them to begin to work with eligible human embryonic stem cell lines); two pilot and feasibility awards; three awards to support exploratory human embryonic stem cell centers; one institutional development award; four post-doctoral training fellowships; one career enhancement award; and six awards to fund stem cell training (including short-term courses) to provide hands-on training to enable researchers to learn the skills and techniques of culturing human embryonic stem cells.

The latter task, of training new researchers, the NIH regards as one of its principal challenges in advancing the field, and, along with available lines and available financial resources, as a key measure of how the field is progressing. As NIH Director Elias A. Zerhouni put it in his presentation before this Council, I don't think the limiting factor is the cell lines. I really don't.

I really think the limiting factor is human capital and trained human capital that can quickly evaluate a wide range of research avenues in stem cells. The NIH has therefore devoted funding to the training of investigators and the cultivation of career development pathways, including short-term courses in stem cell culture techniques and (long-term) career enhancement awards in the field. Some critics have contended, however, that the two issues (funding restrictions and the scarcity of personnel) are likely connected, and that limits on the cell lines eligible for funding and the surrounding political controversy cause some potential researchers to stay away from the field, contributing to a shortage of investigators.25 These federal resources, then, have been directed toward the advancement of human embryonic stem cell research within the bounds of the determination to refrain from supporting or funding new destruction of human embryos.

Scientists may receive federal funding—at any level determined appropriate by the NIH—for any sort of meritorious research, using as many of the approved lines as they are eventually able to use. They can, of course, also receive federal funding for using or deriving new animal embryonic stem cell lines, to assess the potential of these cells for treatment of animal models of human disease (though of course animal models provide only limited information because they are not in many cases exactly extrapolatable to the specific situations that hold in human disease and development, and so cannot replace human cell sources).

Researchers can, in addition, use federal funds for work involving human embryonic germ cells, obtained from aborted fetuses. They can carry out research projects using embryonic germ cell lines already derived, following review and approval of specific institutional assurances, informed consent documents, scientific protocol abstracts, and Institutional Review Board approvals by the NIH's Human Pluripotent Stem Cell Review Group (HPSCRG). They can also receive federal funds for the derivation and study of *new* embryonic germ cell lines following the same HPSCRG review and approval process. In addition, of course, they can develop animal embryonic germ cell lines to assess the potential of these cells through animal models.

Also, researchers can receive federal funds for work conducted on human adult (non-embryonic) stem cells. There are no restrictions regarding what American scientists can do with regard to adult stem cells using taxpayer funds, other than those requiring them to honor the usual human subject protections and clinical research requirements (if they are to be transplanted into human patients). The NIH has devoted substantial resources to the study of human adult stem cells, allocating over $170 million to the field in fiscal year 2002, and approximately $181.5 million in fiscal year 2003 (approximately ten times the amount devoted to human embryonic stem cell work).

Finally, researchers remain free to pursue work (including the derivation of new lines of embryonic stem cells) in the private sector, without government funding. Indeed, as discussed above, embryonic stem cells were first isolated and developed in the private sector, or in university laboratories using private sector funds, and no work in the field was publicly funded at all until 2001. Under present law, work supported by private funds can proceed without restriction.

Under rules promulgated in the spring of 2002, such work does not need to be conducted in a separate laboratory, but a clear separation of the funds used to support this work from any federally funded work of the laboratory is required. Of course, because of the highly interlocking and complex nature of the various aspects of operating a laboratory, such separation can still prove extremely difficult to manage. It is not clear precisely how much privately funded work using human embryonic stem cells has been undertaken in the past few years, but some general figures are available.

The most recent and thorough survey available, based on figures from 2002, suggests that approximately 10 companies in the United States were actively engaged in embryonic stem cell work, employing several hundred researchers and, cumulatively over the past several years, spending over $70 million in the field, which is well over twice what the NIH has so far spent. Those involved in privately-funded research in the field, however, generally do not see private funding as a substitute for federal funds, but would much prefer that the field had the opportunity to benefit from both. They also argue that restrictions on federal funds, and the controversy surrounding the subject, act to dissuade potential investors from entering the field, and thereby have a "chilling effect" on private as well as publicly funded research.

Moreover, just as federal policy can affect privately conducted research, so too a number of states have enacted policies affecting stem cell research, ranging from all-out prohibitions of such research to official statements of support and positive encouragement. The status of such research, and the conditions to which it is subject, can vary dramatically from state to state, independent of federal funding policy.

The administration's policy on the funding of embryonic stem cell research rests on several moral and ethical-legal principles, set upon the reality of existing law:

1. The law: The Dickey Amendment, which the President is required to enforce.

2. The principle underlying the law: The conviction, voiced by the administration, a majority of the Congress, and some portion of the public, that federal taxpayer dollars should not be used to encourage the exploitation or destruction of nascent human life, even if scientific and medical benefits might come from such acts.

3. The principle underlying the desire to offer funding: That efforts to heal the sick and the injured are of great national importance and should be vigorously supported, provided that they respect important moral boundaries.

4. The significance of federal funding: That federal funding constitutes a meaningful positive statement of national approval and encouragement, which should be awarded only with care, particularly in cases where the activity in question arouses significant public moral opposition.

The significance of the policy is best understood in light of these key elements. Its soundness is most reasonably measured against them and against the policy's implementation by the National Institutes of Health.

Though the prudential and principled considerations raised in this chapter governed the formulation of the policy, or at least defined its articulation by its advocates and authors, these are not the only terms by which federal funding policy might be conceived or measured. In this study we present an overview of the ethical and policy debates that have raged for the past two years around both the wisdom of the present policy and the fundamental issues at stake in human embryonic stem cell research.

Appendices

APPENDIX 1

REMARKS BY PRESIDENT GEORGE W. BUSH ON STEM CELL RESEARCH

August 9, 2001

THE PRESIDENT: "Good evening. I appreciate you giving me a few minutes of your time tonight so I can discuss with you a complex and difficult issue, an issue that is one of the most profound of our time.

The issue of research involving stem cells derived from human embryos is increasingly the subject of a national debate and dinner table discussions. The issue is confronted every day in laboratories as scientists ponder the ethical ramifications of their work. It is agonized over by parents and many couples as they try to have children, or to save children already born.

The issue is debated within the church, with people of different faiths, even many of the same faith coming to different conclusions. Many people are finding that the more they know about stem cell research, the less certain they are about the right ethical and moral conclusions.

My administration must decide whether to allow federal funds, your tax dollars, to be used for scientific research on stem cells derived from human embryos. A large number of these embryos already exist. They are the product of a process called in vitro fertilization, which helps so many couples conceive children. When doctors match sperm and egg to create life outside the womb, they usually produce more embryos than are planted in the mother. Once a couple successfully has children, or if they are unsuccessful, the additional embryos remain frozen in laboratories.

Some will not survive during long storage; others are destroyed. A number have been donated to science and used to create privately funded stem cell lines. And a few have been implanted in an adoptive mother and born, and are today healthy children. Based on preliminary work that has been privately funded, scientists believe further research using stem cells offers great promise that could help improve the lives of those who suffer from many terrible diseases — from juvenile diabetes to Alzheimer's, from Parkinson's to spinal cord injuries. And while scientists admit they are not yet certain, they believe stem cells derived from embryos have unique potential.

You should also know that stem cells can be derived from sources other than embryos — from adult cells, from umbilical cords that are discarded after babies are born, from human placenta. And many scientists feel research on these type of stem cells is also promising. Many patients suffering from a range of diseases are already being helped with treatments developed from adult stem cells.

However, most scientists, at least today, believe that research on embryonic stem cells offer the most promise because these cells have the potential to develop in all of the tissues in the body.

Scientists further believe that rapid progress in this research will come only with federal funds. Federal dollars help attract the best and brightest scientists. They ensure new discoveries are widely shared at the largest number of research facilities and that the research is directed toward the greatest public good.

The United States has a long and proud record of leading the world toward advances in science and medicine that improve human life. And the United States has a long and proud record of upholding the highest standards of ethics as we expand the limits of science and knowledge. Research on embryonic stem cells raises profound ethical questions, because extracting the stem cell destroys the embryo, and thus destroys its potential for life. Like a snowflake, each of these embryos is unique, with the unique genetic potential of an individual human being.

As I thought through this issue, I kept returning to two fundamental questions: First, are these frozen embryos human life, and therefore, something precious to be protected? And second, if they're going to be destroyed anyway, shouldn't they be used for a greater good, for research that has the potential to save and improve other lives? I've asked those questions and others of scientists, scholars, bioethicists, religious leaders, doctors, researchers, members of Congress, my Cabinet, and my friends. I have read heartfelt letters from many Americans. I have given this issue a great deal of thought, prayer and considerable reflection. And I have found widespread disagreement.

On the first issue, are these embryos human life — well, one researcher told me he believes this five-day-old cluster of cells is not an embryo, not yet an individual, but a pre-embryo. He argued that it has the potential for life, but it is not a life because it cannot develop on its own.

An ethicist dismissed that as a callous attempt at rationalization. Make no mistake, he told me, that cluster of cells is the same way you and I, and all the rest of us, started our lives. One goes with a heavy heart if we use these, he said, because we are dealing with the seeds of the next generation.

And to the other crucial question, if these are going to be destroyed anyway, why not use them for good purpose — I also found different answers. Many argue these embryos are byproducts of a process that helps create life, and we should allow couples to donate them to science so they can be used for good purpose instead of wasting their potential. Others will argue there's no such thing as excess life, and the fact that a living being is going to die does not justify experimenting on it or exploiting it as a natural resource.

At its core, this issue forces us to confront fundamental questions about the beginnings of life and the ends of science. It lies at a difficult moral intersection, juxtaposing the need to protect life in all its phases with the prospect of saving and improving life in all its stages.

As the discoveries of modern science create tremendous hope, they also lay vast ethical mine fields. As the genius of science extends the horizons of what we can do, we increasingly confront complex questions about what we should do. We have arrived at that brave new world that seemed so distant in 1932, when Aldous Huxley wrote about human beings created in test tubes in what he called a "hatchery." In recent weeks, we learned that scientists have created human embryos in test tubes solely to experiment on them. This is deeply troubling, and a warning sign that should prompt all of us to think through these issues very carefully.

Embryonic stem cell research is at the leading edge of a series of moral hazards. The initial stem cell researcher was at first reluctant to begin his research, fearing it might be used for human cloning. Scientists have already cloned a sheep. Researchers are telling us the next step could be to clone human beings to create individual designer stem cells, essentially to grow another you, to be available in case you need another heart or lung or liver.

I strongly oppose human cloning, as do most Americans.

We recoil at the idea of growing human beings for spare body parts, or creating life for our convenience. And while we must devote enormous energy to conquering disease, it is equally important that we pay attention to the moral concerns raised by the new frontier of human embryo stem cell research. Even the most noble ends do not justify any means.

My position on these issues is shaped by deeply held beliefs. I'm a strong supporter of science and technology, and believe they have the potential for incredible good — to improve lives, to save life, to conquer disease. Research offers hope that millions of our loved ones may be cured of a disease and rid of their suffering. I have friends whose children suffer from juvenile diabetes. Nancy Reagan has written me about President Reagan's struggle with Alzheimer's. My own family has confronted the tragedy of childhood leukemia. And, like all Americans, I have great hope for cures.

I also believe human life is a sacred gift from our Creator. I worry about a culture that devalues life, and believe as your President I have an important obligation to foster and encourage respect for life in America and throughout the world. And while we're all hopeful about the potential of this research, no one can be certain that the science will live up to the hope it has generated.

Eight years ago, scientists believed fetal tissue research offered great hope for cures and treatments — yet, the progress to date has not lived up to its initial expectations. Embryonic stem cell research offers both great promise and great peril. So I have decided we must proceed with great care.

As a result of private research, more than 60 genetically diverse stem cell lines already exist. They were created from embryos that have already been destroyed, and they have the ability to regenerate themselves indefinitely, creating ongoing opportunities for research. I have concluded that we should allow federal funds to be used for research on these existing stem cell lines, where the life and death decision has already been made.

Leading scientists tell me research on these 60 lines has great promise that could lead to breakthrough therapies and cures.

This allows us to explore the promise and potential of stem cell research without crossing a fundamental moral line, by providing taxpayer funding that would sanction or encourage further destruction of human embryos that have at least the potential for life.

I also believe that great scientific progress can be made through aggressive federal funding of research on umbilical cord placenta, adult and animal stem cells which do not involve the same moral dilemma. This year, your government will spend $250 million on this important research.

I will also name a President's council to monitor stem cell research, to recommend appropriate guidelines and regulations, and to consider all of the medical and ethical ramifications of biomedical innovation. This council will consist of leading scientists, doctors, ethicists, lawyers, theologians and others, and will be chaired by Dr. Leon Kass, a leading biomedical ethicist from the University of Chicago.

This council will keep us apprised of new developments and give our nation a forum to continue to discuss and evaluate these important issues. As we go forward, I hope we will always be guided by both intellect and heart, by both our capabilities and our conscience.

I have made this decision with great care, and I pray it is the right one.

Thank you for listening. Good night, and God bless America."

APPENDIX 2

REMARKS BY PRESIDENT GEORGE W. BUSH ON STEM CELL RESEARCH THE MEANING OF FEDERAL FUNDING

I. Introduction

How should the government approach the question of public funding of activities that are deemed controversial by the American people? Is it appropriate to make such decisions on moral grounds? Can moral grounds for such decisions be avoided? If they can't, whose moral views, and of which sort, should govern, and with what consequences for those in the minority? Questions of this sort have been frequently discussed in the wake of President Bush's 2001 decision regarding federal funding of embryonic stem cell research. In that decision, the President permitted federal funds to be used, for the first time, to support research on embryonic stem cells, but only those already in existence. At the same time, he made it clear that there would be no federal support for any research that involved or depended on any future destruction of human embryos. In so doing, he was upholding both the letter and the spirit of a Congressional enactment, the 1996 Dickey Amendment, which prohibited the creation of embryos for use in experiments, or the use of embryos in research that led to their destruction.

President Bush's decision has generated a great deal of controversy. Most scientists and patient advocacy groups believe that he made the wrong decision, and that the Dickey Amendment is a terrible mistake. Among the objections one commonly hears to the President's policy are several that concern the meaning of federal funding: (1) By withholding federal funding for research that involved the creation of *new* embryos or the future destruction of embryos in existence, the President has effectively banned embryonic stem cell research.

(2) The decision was wrong because the President allowed his personal moral views to govern federal policy. Or, along the same lines, the congressional ban is wrong because it represents the imposition of moral views—religiously based moral views at that— to frustrate sound and beneficial public policy.

(3) The decision is morally incoherent, for if an act is so immoral as to deserve the governmental disapproval implicit in withholding funding, it should be accompanied by efforts to prohibit the activity altogether.

Whatever the merits of the current law, or the President's 2001 stem cell decision, these objections, once closely examined, cannot pass muster. The first confuses a limitation on funding with the imposition of a ban or prohibition. The second wrongly supposes that legislating morals through federal budget decisions is always or generally wrong. And the third incorrectly assumes that government has an obligation to bring an end to all conduct it believes immoral.

Explaining these errors requires an exploration of the meaning of all government funding decisions. Such an exploration can not decide the difficult question of the merits of the President's stem cell policy. It can, however, put to rest the objections built around the claim that the policy somehow violates the letter or the spirit of sound constitutional government.

II. Federal Funding

A. Basic Consideration

The common objections to the President's policy fail to come to grips with what government funding in a liberal democracy really means. Several fundamental features of our constitutional system need to be emphasized.

First, no one and no activity has a constitutional right to federal funding. There is no governmental obligation to fund most activities, not even the most worthy, save for such matters as the Constitution explicitly proclaims to be the responsibility of government, such as national defense, the maintenance of federal courts, the holding of elections, and so on. And even concerning these constitutional essentials, it is an open question, to be resolved by our elected representatives, of how government will choose to allocate taxpayer dollars.

Second, no individual or cause has a right to sit at the government trough. Resources are scarce, and insufficient to support all worthy activities. People with different causes and interests compete to obtain them, and in order to succeed they are forced to bring their case to members of Congress. Funds are distributed only through the political process, within limits set by the Constitution, as the result of deliberation, lobbying, deal-making, and the like.

Third, in a healthy democracy people will always have disagreements about what activities should receive government funding. Sometimes the disagreements will be intense, and sometimes not. Sometimes the disagreements will include moral disagreements, and sometimes not. Sometimes the political process will generate a stable compromise on the issue, and sometimes one side or the other decisively prevails.

Fourth, while the Constitution prohibits the government from establishing religion, it does not interfere with the right of citizens to form moral judgments based on their deeply held religious beliefs, and to persuade a majority of their fellow citizens to enact legislation informed by those moral judgments, provided of course that the legislation does not interfere with other constitutionally guaranteed rights.

Fifth, those who fail in the democratic process to obtain federal funding will always feel that they did not get what they need or want, but in the absence of a clear legal entitlement to such funding, they cannot properly complain that the government has thereby denied their rights or interfered with their liberty to exercise them.

Sixth, those who lose have several alternatives built into the democratic process. They can try to persuade their representatives in Congress to reconsider, they can vote in others more sympathetic to their cause, they can seek to influence public opinion, or they can seek non-government funding for their activities.

All of this is straight-forward and uncontroversial. It suggests the legitimacy, indeed the routine character, of the President's policy. It might be regarded as the end of the story.

B. Are There Special Cases?

Although the framework laid out above may correctly describe the situation for most or even all federal funding decisions, there are moral and political reasons why people might regard, for example, withholding of support for selected aspects of biomedical research as a special case, an exception that demands a different approach.

The nation strongly and overwhelmingly backs biomedical research. And we generally leave the mapping of research strategies to scientists and those who administer the institutions in which they work. The entire biomedical enterprise in the US, including also the training of the next generation of scientific researchers, has come to depend heavily on government support. The public generally favors this arrangement, and relies on government-funded research for the treatment and for the cure of all still untreatable diseases, such as cancer and Alzheimer's disease.

Consequently, the decision to withhold public funds from any particular piece of the biomedical research portfolio looks and feels, both to scientists and to the public, like an intrusion of government into a place where it does not belong, and it prompts harsh accusations that government is engaging in censorship or even outright prohibition of medically necessary scientific research. To be sure, the FDA regularly imposes restrictions on research, but mainly on grounds of safety. When, however, government's objection to research is moral in nature, it strikes scientists as a deprivation: a restriction of freedom to inquire, a thwarting of worthy community goals, an intrusion of morals into a sphere where they do not belong.

At the same time, it appears to those members of the public who disagree with the decision as a failure by the government to abide by its putative moral obligation to use its resources to explore *all* fruitful areas of research in search of cures for dread diseases.

Moreover, there is reason to single out for special attention those decisions about federal funding where powerful moral principles are at loggerheads, and the nation is deeply and passionately divided. This is the case of stem cell research. It poses a confrontation between genuine and conflicting goods: on the one side, respect for nascent human life and, on the other side, commitment to unfettered scientific inquiry and to the fight against disabling and deadly disease. The clash between those who hold that the moral status of the embryo is no different from that of a fully developed human being, and those who believe that the embryo is a clump of cells, utterly devoid of moral worth is not resolved by appeal to shared moral premises. This is because what the debate over stem cell policy calls into question is how to *apply* our shared belief in the rights and dignity of the individual.

Despite the powerful presumption in favor of federal funding of biomedical research, and the moral stakes which both sides see as exceedingly high, the controversy over federal funding of stem cell research does not present a special case. In politics, though, how a policy appears is important. For this reason, and despite the fact that the common arguments condemning the president's policy rest on false assumptions or unreasonable expectations, and though they may be mere rationalizations for the failure to win the policy battle, it is worth examining their flaws fully.

II. Morals, Federal Funding, and Legislation

A. Federal Funding

In the first place, federal funding is about resource distribution– who and what will get how much of the nation's scarce taxpayer dollars. It is usually not about restricting basic rights. For example, there is no constitutional right to the funding of biomedical research.

But often the question of whether government will or will not fund an activity is about more than mere distribution. It is about government shaping choices among various and competing goods or undertakings. It is a statement of approval and encouragement by government, a declaration by the nation that an activity or undertaking is meritorious and has priority. Or, in the decision to withhold funds, government policy can be a statement of disapproval and discouragement, a declaration by the nation that a permitted activity or undertaking lacks merit or has low priority.

Policy decisions about funding resemble policy decisions about taxing. Both sorts of decisions create incentives and disincentives by making activities more or less costly. The child tax credit, for example, reduces the financial cost of child rearing. In so doing, it strengthens families in two ways: it enables families to save money, and it conveys an important message about the political importance of the well-being of the family. Similarly, government funding of research into disease and its prevention and treatment increases the supply of these goods, and reflects our nation's considered judgment that the relief of physical suffering is a high national priority.

While all law either requires, forbids, or permits, the provision or withholding of funding and the use of tax incentives and disincentives allows government to express a range of attitudes toward that which it permits. In the

United States, through such decisions government strongly endorses charity and higher education. It looks favorably on national service and the arts. It shows a preference for marriage to cohabitation. It frowns upon smoking. It is the distinction between permitting or tolerating an activity (which is the case with embryo research and destruction) and actively promoting it through governmental funding (which the president's policy on stem cells prohibits), that is crucial to understanding the president's stem cell research policy, but not only to the stem cell controversy.

The question of federal funding routinely implicates questions about the nation's *moral* priorities among permissible activities. And the question of moral priorities in politics is not so simple as a question of good or bad; rather it is a question of better or worse. One consequence of this is that in sorting out funding decisions it is not a matter of one side introducing moral considerations. Most of the time, both sides in disputes over policy are of necessity engaged in making moral arguments.

This is true, and to an extraordinary degree, in the stem cell controversy. *Both* sides–those who wish to defend the rights of nascent human life, and those who wish to defend unfettered scientific research directed at the relief of human suffering through the cure of deadly disease–defend moral principles. To make matters more difficult, *both* sides tend to defend those principles in their absolute form. While the president's position attempts to give weight to *both* sides' principles, *both* sides, because of the passion with which each holds its principle, are dissatisfied. However, insofar as the president's approach reflects that adopted by the Dickey Amendment, it follows a determination made by a majority of the people's representatives, and with that *both* sides should be satisfied. But they aren't.

But the dissatisfaction of both sides takes a recognizable form.

In general, because moral principles are so frequently at stake in the fight for federal taxpayer dollars, funding decisions create bitterness. This truer still, as in the stem cell debate, when the moral principles are wielded in their absolute form, so that *both* sides can claim defeat. If funding is withheld, those who believe the activity is worthy can claim that their tax dollars, which they contribute in the hope that they will serve the good of the country, are being held back from what they deem a deserving or even overriding moral purpose. This is the position taken by many scientists and progressives with regard to the limitations imposed by the President's policy. If funding is provided, those who believe the activity is immoral can claim that their tax dollars are being used to advance a cause they believe is unworthy, or even abhorrent. This is the position taken my some social conservatives who believe that the limitations imposed by the President's policy did not go nearly far enough and pave the way, sooner rather than later to the routine creation and destruction of human embryos for biomedical research.

Typically, both sides make moral claims, and one or the other–and in the really tough cases such as stem cell policy, *both* sides–will have to live with the fact that their moral principles are being rejected (if not assaulted) by the government, in their own name and with their own tax dollars.

Why should those who lose the political struggle put up with this? For the simple reason that living in a liberal democracy means sometimes being in the minority, even on questions of the utmost importance, but so long as the laws which one opposes are consistent with the Constitution and enacted according to legally appropriate procedures, one has an obligation to obey them.

B. Legislation

Is it really a legitimate aim of a liberal democracy to adopt laws and take actions to shape the moral beliefs of its citizens? Perhaps federal funding is the exception, and to the extent possible the moral dimension should be eliminated from policy formation. Doesn't government in a liberal democracy have an obligation to remain neutral toward competing conceptions of a good life, and so refrain from enacting morals into law? Otherwise, doesn't it impermissibly infringe on people's right to choose how to live their lives.

According to a common and sound criticism of this common view of the liberal state, such neutrality is a chimera: it is impossible for any government to remain neutral about morality and the nature of a well-lived life, since the resolution of controversies over public policy–for what purposes is the state permitted to classify citizens by race? what is the meaning of marriage? to what extent may the public schools engage in civic education?– always draw upon, reinforce, or suppress a view about what is deserving, proper and good. It is possible, as a matter of policy, to tolerate a wide variety of choices and forms of life, but toleration itself is a moral principle based on a certain interpretation of how to secure human freedom and respect the dignity of the individual.

Some then object that because of its very foundational commitments, our liberal democracy privileges the autonomous or freely choosing life. And so in sense it does. But it need not and should not do this unwittingly or surreptitiously. The mistake is to think that liberal state stands or falls with the commitment to neutrality. It doesn't. It stands or falls with the commitment to creating the conditions under which individuals can exercise political freedom.

Law and public policy in a liberal democracy properly seek to create conditions in which citizens can make informed and responsible choices. It does this in a variety of ways. The first and most taken for granted is through the establishment of public order.

It also does this through establishing a system of public schools, promoting research in the arts and sciences, and enacting a wide variety of social and economic legislation, all with a view to forming a citizenry that is at home in, and capable of taking advantage of, freedom. Legislation designed to encourage biomedical research can be seen as creating circumstances in which we are better able to enjoy the blessings of freedom. The same can be said of laws designed to respect, and encourage respect, for nascent human life, which can reasonably be understood as contributing to the conditions under which individuals learn to respect humanity in others and in themselves.

To be sure, even within the limits provided by law, government's encouragement of informed and responsible choice can easily become a tool for the ill-conceived circumscribing of choice. Even well meaning government efforts to help prepare citizens for liberty and toleration can undermine both. Government funded education can be dogmatic and ideological; government funded research may be biased and unaccountable; government supported arts may disseminate tawdry or jingoistic sentiments and images; government funded programs directed at the family may fail to adapt to changing times. Of course, these familiar abuses are not arguments against government promoting the conditions that enable citizens to take advantage of freedom. Rather, they are reasons for proceeding with care, and with an appreciation of the complexities of contemporary moral and political life.

IV. American Dilemmas

The president's policy on stem cells is not the only funding decision in contemporary American politics that has generated controversy. Brief discussion of others sheds light on what is common to all and what is distinctive to the stem cell debate.

Consider first the battle over abortion, which involves a long standing struggle over the question of government funding for lawful conduct. Shortly after entering office, President Bush ordered the withholding of funding from international organizations that performed abortions, a decision that was neither required of him nor forbidden to him but within his discretion. The principle behind this policy is common to his position on stem cell research: government funds should not be used to destroy nascent human life.

At home, a line of Supreme Court decisions stretching from 1977 to 1991dealing with abortion and government funding established the principle that the constitution does not require government to fund activities that the Constitution protects. In *Maher v. Roe* 432 U.S.464 (1977), the Court held 6-3 that Connecticut could provide Medicaid benefits for childbirth while withholding benefits from women who wished to have non-medically necessary abortions.

Justice Powell, writing for the majority, maintained that the right to abortion announced in *Roe v. Wade*, "protects the woman from unduly burdensome interference with her freedom to decide whether to terminate her pregnancy. It implies no limitation on the authority of a State to make a value judgment favoring childbirth over abortion, and to implement that judgment by the allocation of public funds." Powell's analysis emphasized the "basic difference between state interference with a protected activity and state encouragement of an alternative." In dissent, Justice Brennan disagreed vociferously. He argued that the denial of funds unconstitutionally interfered with the right of women to choose an abortion. Also in dissent, Justice Marshall argued that by withholding funds for non-medically necessary abortions, Connecticut was seeking "to impose a moral viewpoint that no State may constitutionally enforce." In *Harris v. McRae* 448 U.S. 297 (1980), by a 5-4 margin, the Court upheld the Hyde Amendment, which banned federal funding of medically necessary abortions. The majority's argument was much the same as in *Maher*: "although government may not place obstacles in the path of a woman's exercise of her freedom of choice, it need not remove those not of its own creation." The government, the majority reasoned, does not have an obligation to provide taxpayer dollars so that individuals can exercise their individual rights to the maximum. To this, Justice Brennan replied in dissent that Hyde Amendment actually left poor women in a worse off position. By refusing to provide poor women with funding for even medically necessary abortions while subsidizing childbirth, the government demonstrated profound disapproval for abortion and thereby burdened the exercise of the right to privacy declared in *Roe.*

In *Rust v. Sullivan* 500 U.S. 173 (1991), again by 5-4 margin, the Court upheld federal regulations that barred health care professionals who received federal funding from offering counseling about abortion. Chief Justice Rehnquist, writing for the majority, reiterated the *Maher* and *Harris* principle: "The Government has no constitutional duty to subsidize an activity merely because the activity is constitutionally protected." In dissent, Justice Blackmun, joined by Justices Stevens and Marshall, echoing the dissenters in *Maher* and *Harris,* insisted that by conditioning federal funding on the withholding of counseling about abortion, the government was actually placing "formidable obstacles" in the path of women's exercise of their privacy rights.

While these cases appear closely analogous to the stem cell controversy, they differ in a crucial respect. Although some of the same forces are politically engaged, and although the moral issue concerns the question of the inviolability of nascent human life, the abortion cases would have been unlikely to come to the Court for adjudication had the Court not declared in *Roe v. Wade* a constitutional right to an abortion. For the argument made by those who seek federally funded abortions is that by withholding funding, the government is seeking to frustrate the exercise of a constitutionally protected right. Absent such a right, there could be no plausible legal claim. Indeed, absent an entitlement to government sponsored health care benefits, there is no valid legal claim that Medicare *must* pay for cosmetic surgery, sex-change operations, contraceptive benefits, heart transplants, or any other procedure one wants for oneself and can find a doctor to do. Only if there were a constitutionally protected right not to be poor, not to be without resources to fully take advantage of all the things that we are legally entitled to pursue, could such a claim prevail as a matter of law. While there is no such right, this is just the kind of claim that dissenters in the Court's cases concerning federal funding and abortion defend. In short, because the Constitution provides no special protection to biomedical research, the argument for legal entitlement to funding of stem cell research proceeds on dramatically weaker grounds than the rejected arguments in the abortion funding cases.

Title VI of the Civil Rights Act of 1964 furnishes another example of how government may withhold funds from practices it does not outlaw. It provides that, "No person in the United States shall, on the ground of race, color, or national origin, be excluded from participation in, be denied the benefits of, or be subjected to discrimination under any program or activity receiving Federal financial assistance." It is this provision that requires private universities to avoid those racial classification in admissions and hiring that would violate the prohibitions imposed on *state* action by the equal protection clause of the 14th Amendment. Title VI is far reaching, because most private universities rely heavily on government funding for the support of basic research. And it provides a way for the federal government to shape the moral contours of what is largely private conduct, and bring that conduct in line with fundamental constitutional principles. Of course private institutions are free to continue to practice activities that disqualify them for federal funding. All they have to do is refuse to take federal funds.

Close in form to federal policy on stem cell research are social security regulations regarding marriage and survivor benefits. For example, although cohabitation without matrimony is not illegal, indeed it is quite common, the federal government refuses to pay social security survivor benefits to all but legal spouses. This is a way for government to provide financial incentives for marriage.

And for government to take sides on the value of marriage, proclaiming the union marked by it as good for individuals and good for the polity. This is a policy decision that does not engender bitterness or controversy. It must be acknowledged that the withholding of a reward could, under imaginable circumstances, stigmatize those who choose to live together as a loving couple but not to marry. But just as it can not plausibly be claimed today that the child tax credit confers social disapprobation on married couples without children, so too it cannot be plausibly claimed that unmarried couples suffer social disapprobation because of government policy that restricts the paying of social security survivor benefits to legal spouses.

Or, from a different angle, consider the question of elementary level and high school education. In 1923, in a landmark decision, *Meyer v. Nebraska* 262 U.S. 390 (1923), the Supreme Court ruled that parents have a right to educate their children in a foreign language.

In 1925, in a related case, *Pierce v. Society of Sisters* 268 U.S. 510 (1925), the Supreme Court ruled that parents have a right to educate their children in private schools. But nobody concludes that the rights that these cases protect *prohibit* states from policy decisions encouraging public education. And nobody claims that the right of parents to privately educate their children creates an *entitlement* to have that private education funded by the government.

As these examples illustrate, the controversy over stem cells should be seen as one among many political battles over the allocation of limited federal funds. It is distinguished not by the presence of moral principles, or the presence of moral principles on both sides, but by the particular moral principles at stake, the absoluteness with which they are wielded, and the intensity of the passions their defense provokes.

When the question of federal funding is placed in perspective,it can be seen that the common objections to the President's policy on stem cell research are misplaced.

First, by withholding federal funding for research that involved the creation of new embryos or the future destruction of embryos, the President did not effectively *ban* embryonic stem cell research. His August 2001 decision for the first time provided federal funding for stem cell research, and it permitted private individuals and companies to pursue it. Second, by basing his policy in part on moral considerations, the President did not violate an obligation to keep morals out of politics, because funding decisions, whichever way they go, typically and unavoidably contain a moral component. Indeed, the moral component often lies at the heart of the dispute and at the core of the decision. Third, by refusing to seek a blanket prohibition on an activity from which he withheld funding on moral grounds, the President did not make an incoherent decision. The complexities of a free society frequently create situations in which it makes sense for government to express doubt, anxiety or disapproval for an activity it is unwilling or unable to outlaw.

None of this is to deny that the president's policy on stem cell research is open to criticism on the merits. It is only to claim that the policy reflects a perfectly appropriate exercise of governmental powers in a liberal democracy.

APPENDIX 3

HUMAN EMBRYONIC GERM CELLS

June 2001 – July 2003

There have been but two original research articles published on human embryonic germ cell in the period covered by this report. It is appropriate that only peer-reviewed articles be considered in this report, as the field of stem cell research is rife with undocumented or unsubstantiated claims. There have been several publications on EG cells derived in mice and chick and comment will be made on these reports as they may impact on the eventual use or study of the human cells.

There is a concern, given the imprinting-based developmental abnormalities observed in humans, and those that have been produced experimentally in animals, that dysregulated gene expression in stem cell derived tissues could pose a serious problem in the use of such tissues for cell-based therapies. Genomic imprinting is defined as an epigenetic modification of the DNA (other than sequence) in the germ line that leads to the preferential expression of a specific allele (monoallelic expression) of some genes in the somatic cells of the offspring in a parental dependent manner.

Because a maternal gene allele may be a paternal allele in the next generation, the imprinting must be reprogrammed in the germ line, that is, the epigenetic 'marks' must be erased (during germ cell development) and established in the newly formed embryo. The timing of the erasure in humans is not known. Imprinting involves methylation at specific sites in the DNA and most likely, other changes as well.

There have been several reports from experiments with the mouse that indicate that imprinting is abnormal in pluripotent stem cells derived from mouse embryos. These studies include abnormal or variable imprints in mouse EG cells, abnormal imprints in ES cells derived from interspecific crosses, and abnormal gene expression in mice derived from ES cell nuclear transfer.

The results of the study by Onyango *et al.* utilizing EG cell lines derived at John Hopkins, clearly demonstrate that general dysregulation of imprinted genes will not be a barrier to their use in transplantation therapies. The report has determined that the EG cells are not imprinted, that is, imprinting has been erased in the primordial germ cells that gave rise to the EG cells, that the erasure is maintained in the EG cells, but in all informative cases, they observed the transcription of only a single allele in differentiated cells derived from the human EG cells. These results, although on a limited number of lines and only a few imprinted genes, would indicate that these human EG cell lines will serve as reliable and safe sources for the study of EG cell differentiation and, perhaps, cells for cell-based interventions.

Another area of interest in genetic regulation within EG cells is that of X inactivation, the mammalian method for equalization of the dosage of X-linked genes in males and females. This equalization is accomplished by the down regulation of the transcriptional output of the X chromosomes in females, so that only one X is active in diploid somatic cells of both sexes. Inactivation is initiated in female blastocysts. Both X chromosomes in female primordial germ cells are active. Migeon *et al.* (2001) have demonstrated that in the very early stages of differentiation of cells from human EGs, only one X chromosome is active, indicating normal genetic regulation has occurred. In a report by Nesterova *et al.* (2002) on the use of mouse female EG cells in the study of X chromosome inactivation/ reactivation during primordial germ cell migration and EG cell formation. Both X chromosomes appear to be active in XX EG cells, and presumably, one becomes inactive when cells differentiate from the EG cells.

One of the goals of stem cell research is to provide sources of cells for cell-based therapies. As a step in this direction, proof of concept or proof of principle studies involve the use of human cells in animal models of human disease or injury. Although few, if any, animal models are true models for the human diseases, they are the closest approximation that can be made. The first report on the use of cells derived from stem cells of human embryonic sources was recently published: Kerr *et al.* Human Embryonic

Germ Cell Derivatives Facilitate Motor Recovery of Rats with Diffuse Motor Neuron Injury, J. Neuroscience 23, June 15, 2003. This is the first demonstration that a human pluripotent stem cell derived form embryonic or fetal tissue can ameliorate a disease process in an animal model.

Neural progenitor cells, derived from human EG cells, were introduced into the cerebrospinal spinal fluid of rats that had been paralyzed as a result of infection with a neuroadapted Sindbis virus that specifically targets motor neurons in the spinal cord. All animals in which human cells were found had some degree of hindlimb recovery. It was clear from the histology of the animals that the human cells had differentiated into appropriate neural cell types within the ventral horns, including motor neurons, the results indicated that the major effect of the human cells was to protect host neurons from death and to facilitate reafferentiation of motor neuron cell bodies. Growth factors responsible for this recovery, produced by the human cells, were identified as brain-derived neurotrophic factor and transforming growth factor.

The significance of this experiment appears to be that for this disease model, the human cells supply factors that facilitate motor neuron recovery following viral damage. However, the cells did migrate to the site of injury in the spinal cord, which differentiates them from other cellular grafts, and then delivered the factors. Also, there was considerable differentiation of the human cells into various neural cells types in the cord.

In a review article on deriving glucose-responsive insulin-producing cells from stem cells (Kaczorowski *et al.* 2002), mention is made on the isolation of such cells from mouse and human ES cells and human EG cells, but no new data was presented, only a reference to a paper published in 2001.

EG cell studies in other species

Several studies on EG cells from other species have been published.

EG cells from the chick have been demonstrated to yield germ line chimeras when transferred to early embryos (Park *et al.* 2003). EG cell lines of the mouse were found to colonize not only the epiblast but also the primary endoderm of the gastrulating embryo following aggregation with 8-cell embryos (Durcova-Hills *et al.* 2003). This observation was permitted as a result of using a transgenic construct effect as a lineage marker for cells of the early embryo. Horii *et al.*

(2002) report the serum-free culture of mouse EG cells, a system which inhibits the 'spontaneous' differentiation due to the presence of various growth factors in the serum. The cells are, however, are cultured on a feeder layer.

An area of great interest in EG cell derivation is the basis of the underlying mechanism for the conversion, derivation or transformation (terms reflecting our ignorance of the process) of primordial germ cells to EG cells. Kimura *et al.* (2003) report that the loss of a tumor suppressor gene in mice, PTEN (phosphatase and tensin homology deleted form chromosome ten, aka MMAC1 and TEP1), leads to a high incidence of testicular teratomas and enhances the production of EG cells from PGCs of the mutant mice.

While it is highly unlikely that the loss of this gene in the culture of normal PGCs leads to the derivation of EG cells, it may implicate downstream signaling pathways that are involved in cell cycle progression, cell survival and cell migration as important players in the process.

APPENDIX 4

MULTIPOTENT ADULT PROGENITOR CELLS: AN UPDATE

Phenotype of Bone Marrow MAPC: MAPC can be cultured from human, mouse and rat bone marrow (BM). Unlike MSC, MAPC do not express major histocompatibiliy (MHC)-class I antigens, do not express, or express only low levels of, the CD44 antigen, and are CD105 (also endoglin, or SH2) negative . Unlike hematopoietic stem cells (HSC), MAPC do not express CD45, CD34, and cKit , but like HSC, MAPC express Thy1, AC133 (human MAPC) and Sca1 (mouse) albeit at low levels .In the mouse, MAPC express low levels of stage specific embryonic antigen (SSEA)-1, and express low levels of the transcription factors Oct4 and Rex1, known to be important for maintaining embryonic stem (ES) cells undifferentiated and to be down-regulated when ES cells undergo somatic cell commitment and differentiation .

MAPC can also be isolated from other tissues, and other species: We also showed that MAPC can be cultured from mouse brain and mouse muscle . Of note, the differentiation potential and expressed gene profile of MAPC derived from the different tissues appears to be highly similar. These studies used whole brain and muscle tissue as the initiating cell population, therefore containing more than neural cells and muscle cells, respectively. The implications of this will be discussed below. Studies are ongoing to determine if cultivation of MAPC from other organs is possible, and whether culture of MAPC, like ES cells, is mouse-strain dependent.

Initial studies suggest that a population of MAPC-like cells can also be cultured from bone marrow from cynomologous monkeys (unpublished observations)(studies done by our collaborator Felipe Prosper, University of Navarra, Pamplona, Spain) and from bone marrow of dogs (unpublished observations)(studies done at the University of Minnesota).

Non-senescent nature of MAPC: Unlike most adult somatic stem cells, MAPC proliferate without obvious signs of senescence, and have active telomerase.

In humans, the length of MAPC telomeres is 3-5kB longer than in neutrophils and lymphocytes, and telomere length is not different when MAPC are derived from young or old donors . This suggests that MAPC are derived from a population of cells that either has active telomerase *in vivo*, or that is highly quiescent *in vivo*, and therefore has not yet incurred telomere shortening *in vivo*. In human MAPC cultures we have not yet seen cytogenetic abnormalities. As human MAPC are however undergoing symmetrical cell divisions, it remains possible that despite lack of gross cytogentic changes, minor mutations accumulate over time. We are therefore planning to use comparative genomic hybridization to address the question at what time genetic abnormalities occur, if they do. Initial results from gene array analysis suggest that MAPC, like ES cells, have a large number of DNA repair genes expressed (unpublished observations), which may protect them from more frequent genetic abnormalities in view of the fact that they undergo multiple sequential symmetrical cell divisions.

However, several subpopulations of mouse MAPC, and to a lesser extent rat MAPC, have become aneuploid, even though additional subpopulations thawed subsequently were cytogenetically normal. Aneuploidy is seen more frequently once mouse (and rat) MAPC have been expanded for >60-70 population doublings and following repeated cryopreservtions and thawing episodes. This characteristic of mouse MAPC is not dissimilar from other mouse cell populations, including mouse ES cells.

Stringent culture conditions required for maintenance of the undifferentiated state of MAPC: Culture of MAPC is, however, technically demanding. Major factors that play a role in successful maintenance of MAPC include cell density, CO concentration and pH of the medium, lot of fetal calf serum that is used, and even the type of culture plastic that is used. Control of cell density appears to be species specific: mouse, rat and perhaps cynomologous monkey MAPC need to be maintained at densities between 500 and 1,000 cells / cm2, whereas human and perhaps dog MAPC need to be maintained between 1,500 and 3,000 cells/ cm . The reason why MAPC tend to differentiate to the default MSC lineage when maintained at higher densities is not known. However, for MAPC to have clinical relevance, this will need to be overcome. Gene array and proteomics studies are ongoing to identify the contact and / or soluble factors that may be responsible for causing differentiation when MAPC are maintained at higher densities. These very demanding technical skills can however be "exported" from the University of Minnesota as, after training at the University of Minnesota, investigators at the University of Tokai, Japan (manuscript submitted) and investigators at the University of Gent, Belgium have successfully isolated MAPC from human bone marrow, and investigators at the University of Navarra, Spain, have successfully isolated MAPC from rat bone marrow.

In Vitro Differentiation Potential of MAPC

We published last year that human, mouse and rat MAPC can be successfully differentiated into typical mesenchymal lineage cells, including osteoblasts, chondroblasts, adipocytes and skeletal myoblasts . In addition, human, mouse and rat MAPC can be induced to differentiate into cells with morphological, phenotypic and functional characteristics of endothelial cells , and morphological, phenotypic and functional characteristics of hepatocytes .

Neuroectodermal differentiation: Since then, we have also been able to induce differentiation of MAPC from mouse bone marrow into cells with morphological, phenotypic and functional characteristics of neuroectodermal cells .

Differentiation of MAPC to cells with neuroectodermal characteristics occurred by initial culture in the presence of basic fibroblast growth factor (bFGF) as the sole cytokine, followed by culture with FGF-8b and sonic hedgehog (SHH), and then brain derived neurotrophic factor (BDNF) . Differentiation using these sequential cytokine stimuli was associated with activation of transcription factors known to be important in neural commitment *in vivo* and differentiation from NSC and mES cells *in vitro*. Cells staining positive for astrocyte, oligodendrocyte and neuronal markers were detected. Neuron-like cells became polarized, and as has been described in most studies in which ES cells or NSC were differentiated *in vitro* to a mid-brain neuroectodermal fate using FGF8 and SHH, approximately 25% of cells stained positive for dopaminergic markers, 25% for serotonergic markers, and 50% for GABA-ergic markers. Subsequent addition of astrocytes induced further maturation and prolonged survival of the MAPC-derived neuron-like cells, which now also acquired electrophysiological characteristics consistent with neurons, namely voltage gated sodium channels and synaptic potentials .

Muscle differentiation: In addition, we now have convincing evidence that MAPC can differentiate into cells with phenotypic as well as functional characteristics of smooth muscle cells (manuscript in preparation).

Interestingly, the lineage that continues to be elusive is cardiac myoblasts, despite the fact that mouse MAPC injected in the blastocyst contribute to the cardiac muscle . Although a number of *in vitro* differentiation conditions induce expression of Nkx2.5, GATA4, and myosin heavy chain mRNA and proteins , we have been unable to induce differentiation of MAPC to cells with the typical functional characteristic of cardiac myoblasts, i.e.

spontaneous rhythmic contractions or beating, a differentiation path that is almost a default differentiation pathway for mouse ES cells.

The reason for the lack of functional cardiac myoblast properties is currently unknown.

Another important cell lineage that has not yet been generated is insulin-producing cells, even though initial studies suggest that differentiation to cells expressing at least early pancreatic and endocrine pancreas transcription factors can be obtained.

In vitro differentiation of MAPC as model system for gene discovery: A last comment regarding *in vitro* differentiation of MAPC is that, in contrast to differentiation of ES cells *in vitro*, the final differentiated cell product derived from MAPC is commonly >70- 80% pure. This should allow using these *in vitro* differentiation models for gene and drug discovery. For instance, in a recently published study we compared the expressed gene profile in human MAPC induced to differentiate to osteoblasts and chondroblasts, two closely related cell lineages. We could demonstrate that although a large number of genes are co-regulated when MAPC differentiate to these two lineages, specificity in differentiation can readily be detected. For instance a number of known and yet to be fully characterized transcription factor mRNAs were differentially expressed during the initial phases of differentiation. Studies are ongoing to further define the role of these genes in lineage specific differentiation. These studies exemplify however the power of this model system to study lineage specific differentiation *in vitro*.

Degree of Pluripotency of MAPC

We have shown that transfer of 10-12 mouse MAPC into mouse blastocysts results in the generation of chimeric mice. When 10-12 MAPC, expanded for 50-55 population doublings, were injected approximately 80% of offspring were chimeric, with the degree of chimerisms varying between 1–40% . Cells found in different organs acquire phenotypic characteristics of the tissue. For instance MAPC derived cells detected in the brain of chimeric animals differentiate appropriately into region specific neurons, as well as astrocytes and oligodendrocytes . More recent studies using MAPC from later population doublings have shown that the frequency of chimerism decreases when MAPC are maintained for longer time in culture, even though animals with chimerism of more than 70% could be obtained (unpublished observations). These studies indicate that like ES cells, MAPC can give rise to most if not all somatic cell types of the mouse. Whether MAPC can do this without help of other cells in the inner cell mass, i.e. can generate a mouse by tetraploid complementation , is not yet known. Also not yet known is whether MAPC contribute to the germ line when injected in the blastocyst.

Post-Natal Contribution to Tissue

Neither human nor mouse MAPC injected into the muscles of severe combined immunodeficient (SCID) mice have led to the development of teratomas (unpublished observations). Likewise, we have not yet detected donor-derived tumor formation following IV injection of human or mouse MAPC in NOD-SCID animals.

However, when mouse undifferentiated MAPC are administered IV to NOD-SCID mice, engraftment in the hematopoietic system as well as epithelia of gut, liver and lung is seen . Preliminary studies using human MAPC suggest that a similar pattern of engraftment may occur, even though the level of contribution to blood, liver, gut and lung is lower (unpublished observations). Noteworthy is the fact that neither mouse nor human MAPC appear to contribute to other tissues when injected IV, except to endothelium (see below).

Although PCR analysis for human DNA in human – mouse transplants or for b-galactosidase in mouse-mouse transplants yielded positive signals in many tissues, we believe that this is mainly due to contaminating blood cells. When tissues were carefully examined for tissue specific differentiated MAPC progeny, we could not detect MAPC-progeny in brain, skeletal muscle, cardiac muscle, skin or kidneys. Lack of engraftment in brain, skeletal and cardiac muscle may be due to the fact that transplants were done in noninjured animals, where the blood brain barrier is intact, and where little or no cell turnover is expected in muscle. More difficult to explain is the absence of MAPC-derived progeny in skin, possibly the organ with the greatest cell turnover. Studies are ongoing to trace the homing behavior of MAPC following infusion in non-injured animals and injured animals, which may shed light on these observations.

In vivo differentiation into skeletal muscle: Muguruma et al have also shown that undifferentiated human MAPC injected in the muscle of non-obese diabetic (NOD)- SCID mice differentiate into cells that stain positive for muscle transcription factors and muscle cytoskeletal proteins (manuscript submitted). Similar results were seen in Minnesota. We also found that pre-treatment of human MAPC with 5-azacytidine, required to induce muscle differentiation *in vitro*, enhanced the degree of engraftment of human cells in mouse muscle, suggesting that predifferentiation of MAPC may under certain circumstances enhance the level of engraftment (unpublished observations).

Contribution to endothelium in vivo: When endothelial cells generated from human MAPC by incubation *in vitro* with vascular endothelial growth factor (VEGF) were infused in animals in which a tumor had been implanted underneath the skin, we detected enhanced tumor growth and found that up to 30% of the tumor vasculature was derived from the human endothelial cells. Likewise, wounds in the ears of these animals as a result of ear tagging contained human endothelial cells. One of the animals developed a host-tumor, an occurrence seen frequently in aging NOD-SCID mice. We detected contribution of MAPC-derived endothelium to tumor vessels . Likewise, one of the NOD-SCID mice that received human MAPC developed a host thymic lymphoma. Human MAPC, like mouse MAPC, appeared to differentiate into endothelial cells that contribute to tumor angiogenesis.

Engraftment of MAPC in stroke model: In yet another *in vivo* study we evaluated the effect of human MAPC in a rat stroke model. Cortical brain ischemia was produced in male rats by permanently ligating the right middle cerebral artery distal to the striatal branch. Animals were placed on cyclosporine-A and 2 weeks later, 2x10 human MAPC were injected around the infarct zone. As controls, animals received normal saline or MAPC conditioned medium. Limb placement test and tactile stimulation test were blindly assessed 1 week before brain ischemia, 1 day before transplantation, and at 2 and 6 weeks after grafting. The limb placement test included eight subtests described by Johansson and coworkers . In a tactile stimulation test , a small piece of adhesive tape was rapidly applied to the radial aspect of each forepaw. The rats were then returned to their home cages, and the order of the tape removal (i.e., left versus right) was recorded. Three to five trials were conducted on each test day.

Each trial was terminated when the tapes were removed from both forepaws or after 3 min. Animals were subsequently sacrificed to determine the fate of the human cells injected in the brain. After 2 and 6 weeks, animals that received human MAPC scored statistically significantly better in the limb placement test as well as tactile stimulation test compared with animals that received only cyclosporine-A (CSA), or were injected with normal saline or MAPC conditioned medium. The level of recuperation of motor and sensory function was 80% of animals without stroke. When the brain was examined for the presence and differentiation of human MAPC to neuroectodermal cells, we found that human MAPC were present, but remained rather immature. Therefore, we cannot attribute the motor and sensory improvement to region specific differentiation to neuronal cells and integration of neurons derived from MAPC in the host brain. Rather the improvement must be caused by trophic effects emanated by the human MAPC to either improve vascularization of the ischemic area, to support survival of the remaining endogenous neurons, or to recruit neuronal progenitors from the host brain. These possibilities are currently being evaluated.

Possible Mechanisms Underlying the Phenomenon of Multipotent Adult Progenitor Cells

Currently we do not fully understand the mechanism(s) underlying the culture selection of MAPC. We have definitive data to demonstrate that the pluripotency of MAPC is not due to coculture of several stem cells.

Pluripotency cannot be attributed to multiple stem cells: First, using retroviral marking studies we have definitive proof that a single cell can differentiate *in vitro* to cells of mesoderm, both mesenchymal and non-mesenchymal, neuroectoderm and hepatocyte-like cells, and this for human , mouse and rat MAPC . Second, we have shown that a single mouse MAPC is sufficient for generation of chimeric animals . Indeed, we published that 1/3 animals born from blastocysts in which a single MAPC was injected were chimeric with chimerism degrees varying between 1 and 45%.

This rules therefore out that the pluripotent nature of these cells is due to co-existence in culture of multiple somatic stem cells.

Cell fusion is not likely explanation: A second possibility for the greater degree of differentiation potential would be that cells undergo fusion and acquire via this mechanism greater pluripotency. Fusion has been shown to be responsible for apparent ES characteristics of marrow and neural stem cells that had been cocultured with ES cells *in vitro*, and more recently for the apparent lineage switch of bone marrow cells to hepatocytes when hematopoietic cells were infused in animals with hereditary tyrosinemia due to lack of the fumarylacetoacetate hydroxylase (FAH) gene [Wang et al, Nature 2003]. In the former two studies, the majority of genes expressed in the marrow or neural cell that fused with ES cells were silenced, and the majority of the genes expressed in ES cell were persistently expressed. Likewise for the bone marrow-hepatocyte fusion, the majority of genes expressed normally in hematopoioetic cells (except the FAH gene) were silenced whereas genes expressed in hepatocytes predominated. Finally, the cells generated were in general tetraploid or aneuploid.

We do not believe that this phenomenon underlies the observation that MAPC are pluripotent. Cultivation and differentiation *in vitro* (in general, except the final differentiation step for neuroectoderm) does not require that MAPC are co-cultured with other cells, making the likelihood that MAPC are the result of fusion very low. Smith et al suggested in a recent commentary that MAPC could be caused by fusion of multiple cell types early on during culture leading to reprogramming of the genetic information and pluripotency. However, we have no evidence that MAPC are tetraploid or aneuploid early during culture, making this possibility less likely. Nevertheless, studies are ongoing to rule this out. The in vivo studies were not set up to fully be capable of ruling out this possibility.

However, a number of findings suggest that fusion may not likely be the cause for the engraftment seen postnatally, nor the chimerism in the blastocyst injection experiment. The frequency of the fusion event described for the ES-BM, ES-NSC, and HSChepatocyte fusion was in general very low, i.e. 1/100,000 cells.

Expansion of such fused cells could only be detected when drug selection was applied in the in vitro systems, and withdrawal of NTBC (2-(2-nitro-4-trifluoro-methylbenzoyl)-1,3-cyclohexanedione) in the FAH mouse model was used to select for cells expressing the FAH gene. The percent engraftment seen in our post-natal transplant models was in the range of 1% - 9%. The frequency of chimerism seen in blastocyst injection studies ranged between 33% and 80% when 1 and 1 and 10-12 MAPC were injected, respectively. These frequencies are significantly higher than what has been described for fusion events with ES cells *in vitro,* and in the HSC-hepatocyte fusion studies *in vivo.*

Furthermore, in contrast to what was described in the papers indicating that fusion may be responsible for apparent plasticity, all *in vivo* studies done with MAPC were done without selectable pressure, mainly in non-injured animals. Therefore, it is less likely that the pluripotent behavior of MAPC in vivo is due to fusion between the MAPC and the tissues where they engraft / contribute to. However, specific studies are currently being designed to formally rule this out.

Primitive ES-like cells that persist vs. de-differentiation: Currently, we do not have proof that MAPC exist as such *in vivo*. Until we have positive selectable markers for MAPC, this question will be difficult to answer. If the cell exists *in vivo*, one might hypothesize that it is derived for instance from primordial germ cells that migrated aberrantly to tissues outside the gonads during development. It is, however, also possible that removal of certain (stem) cells from their *in vivo* environment results in "reprogramming" of the cell to acquire greater pluripotency. The studies on human MAPC suggest that such a cell that might undergo a degree of reprogramming is likely a protected (stem) cell *in vivo*, as telomere length of MAPC from younger and older donors is similar, and significantly longer than what is found in hematopoietic cells from the same donor. The fact that MAPC can be isolated from multiple tissues might argue that stem cells from each tissue might be able to be reprogrammed. However, as was indicated above, the studies in which different organs were used as the initiating cell population for generation of MAPC did not purify tissue specific cells or stem cells. Therefore, an alternative explanation is that the same cells isolated from bone marrow that can give rise to MAPC in culture might circulate, and be collected from other organs. However, we have until now been unsuccessful in isolating MAPC from blood or from umbilical cord blood, arguing against this phenomenon.

Finally, cells selected from the different organs could be the same cells resident in multiple organs, such as MSC that are present in different locations, or cells associated with tissues present in all organs such as for instance blood vessels. Studies are ongoing to determine which of these many possibilities is correct.

We believe that MAPC would have clinical relevance whether they exist *in vivo*, or are created *in vitro*. However, understanding the nature of the cell will have impact on how one would approach their clinical use. If they exist *in vivo*, it will be important to learn where they are located, and to determine whether their migration, expansion and differentiation in a tissue specific manner can be induced and controlled *in vivo*. If they are a culture creation, understanding the mechanism underlying the reprogramming event will be important as that might allow this phenomenon to happen on a more routine and controlled basis.

Either way, a long road lies ahead before MAPC might be applicable in clinical trials. Hurdles to be overcome include development of robust culture systems that will allow automatization. Like of other stem cells, including ES cells, we will need to determine in preclinical models whether undifferentiated vs. lineage committed vs. terminally differentiated cells should be used to treat a variety of disorders. If lineage committed or terminally differentiated cells will be needed, robust clinical scale differentiation cultures will need to be developed. Furthermore, studies will need to be performed to demonstrate whether potentially contaminating undifferentiated MAPC will interfere with engraftment, and / or differentiate inappropriately in vivo. Likewise, studies aimed at determining what level of HLA-mismatch will be tolerated in transplantations, whether tolerization via hematopoietic engraftment from MAPC will be required. As is also the case for other extensively cultured cells, we will need to further determine if prolonged expansion leads to genetic abnormalities in cells that might lead to malignancies when transplanted in vivo.

As a final remark, MAPC appear to have pluripotent potential both *in vitro* and *in vivo*. Furthermore, they appear to proliferate without obvious senescence when maintained under very stringently controlled culture conditions. Because of these reasons, some have argued that they might be a viable alternative to ES cells.

However, at this stage of the research, I feel that such a conclusion is premature. Whether MAPC have equal longevity as ES cells, and have the ability to create all >200 cell types in the body is still not known. Moreover, there appear to be certain cell types that are more readily generated from ES cells compared with MAPC, such as for instance cardiac myoblasts, whereas it appears for instance more easy to generate hepatocyte like cells from MAPC than ES cells.

Therefore, I continue to strongly believe that strict comparative studies between the two cell populations are needed to determine the true potential of the cells, and that the scientific insights gained from these studies should be used to determine which of the cells will be suitable for use in the clinical setting.

APPENDIX 5

ADULT STEM CELLS

Within just a few years, the possibility that the human body contains cells that can repair and regenerate damaged and diseased tissue has gone from an unlikely proposition to a virtual certainty.

Adult stem cells have been isolated from numerous adult tissues, umbilical cord, and other non-embryonic sources, and have demonstrated a surprising ability for transformation into other tissue and cell types and for repair of damaged tissues. This paper will examine the published literature regarding the identity of adult stem cells and possible mechanisms for their observed differentiation into tissue types other than their tissue of origin. Reported data from both human and animal studies will be presented on the various tissue sources of adult stem cells and the differentiation and repair abilities for each source, especially with regards to current and potential therapeutic treatments.

Adult stem cells have received intense scrutiny over the past few years due to surprising discoveries regarding heretofore unknown abilities to form multiple cell and tissue types, as well as the discovery of such cells in an increasing number of tissues. The term "adult stem cell" is somewhat of a misnomer, because the cells are present even in infants and similar cells exist in umbilical cord and placenta. More accurate terms have been proposed, such as tissue stem cells, somatic stem cells, or post-natal stem cells.

However, because of common usage this review will continue to use the term adult stem cell.

This paper will review the literature related to adult

stem cells, including current and potential clinical applications (with apologies to the many who are not cited, due to the exponential increase in papers regarding adult stem cells and the limitations of this review.) The focus will be on human adult stem cells, but will also include results from animal studies which bear on the potential of adult stem cells to be used therapeutically for patients.

This paper will not attempt to review the literature related to hematopoietic stem cells, *i.e.*, the bone marrow stem cell that is the immediate precursor for blood cells, and the formation of typical blood cells. Nor will this paper review the substantial literature regarding clinical use of bone marrow or bone marrow stem cell transplants for hematopoietic conditions such as various cancers and anemias, nor the striking clinical results seen for conditions such as scleromyxedema, multiple sclerosis, systemic lupus, arthritis, Crohn's disease, etc. In these instances, the stem cells are used primarily to replace the hematopoietic system of the patient, after ablation of the patient's own bone marrow hematopoietic system. Finally, multipotent adult progenitor cells (MAPC's), a bone marrow stem cell that has shown significant abilities at proliferation in culture and differentiation into other body tissues, have been reviewed by Dr. Catherine Verfaillie in a separate paper for the President's Council on Bioethics, and the reader is directed to that review for more information.

Key questions regarding adult stem cells are: (1) their identity, (2) their tissue source of origin, (3) their ability to form other cell or tissue types, and (4) the mechanisms behind such changes in differentiation and effects on tissues and organs. Historically only a few stem cells were recognized in humans, such as the hematopoietic stem cell which produces all of the blood cell types, the gastrointestinal stem cell associated with regeneration of the gastrointestinal lining, the stem cell responsible for the epidermal layer of skin, and germ cell precursors (in the adult human, the spermatogonial stem cell.) These stem cells were considered to have very limited repertoires, related to replenishment of cells within their tissue of origin. These limitations were considered to be a normal part of the developmental paradigm in which cells become more and more restricted in their lineage capabilities, leading to defined and specific differentiated cells in body tissues. Thus, discovery of stem cells in other tissues, or with the ability to cross typical lineage boundaries, is both exciting and confusing because such evidence challenges the canonical developmental paradigm.

Stem Cell Markers

Identification of cells typically relies on use of cell surface markers—cellular differentiation (CD) antigens—that denote the expression of particular proteins associated with genomic activity related to a particular differentiation state of the cell. Identification also has relied on morphological and molecular indications of function, such as expression of specific enzymes. Since stem cells by definition have not yet taken on a specific differentiated function, their identification has relied primarily on use of cell surface markers, and only secondarily on production of differentiated products in various tissues. One stated goal has been to isolate a single putative adult stem cell, characterized fully by specific markers and molecular characteristics, and then to follow the differentiation of this single cell (and/or its progeny) to show that it indeed has multipotent or pluripotent capabilities (clonogenic ability). For bone marrow stem cells, selection of putative adult stem cells has usually excluded typical markers for hematopoietic lineages (lin), CD45, CD38, with inclusion or exclusion of the hematopoietic marker CD34 and inclusion of the marker c-kit (CD117). Other proposed markers for adult stem cells are AC133-2 (CD133), which is found on many stem cell populations, and C1qR , the receptor for complement molecule C1q, found on a subset of CD34 human stem cells from bone marrow and umbilical cord blood. When transplanted into immunodeficient mice, C1qR -positive human stem cells formed not only hematopoietic cells but also human hepatocytes. Other methods of isolation and identification include the ability of putative stem cells to exclude fluorescent dyes (rhodamine 123, Hoechst 33342), allowing isolation by fluorescence-activated cell sorter (FACS) of a "side population" of cells within a tissue that have stem cell characteristics. Expression of the *Bcrp1* gene (ABCG2 gene in humans) is apparently responsible for this dye exclusion, and could provide a common molecular expression marker for stem cells . A study of expressed genes from a single cell-derived colony of human mesenchymal stem cells identified transcripts from numerous cell lineages, and a similar attempt at profiling the gene expression of human neural stem cell in culture with leukemia inhibitory factor (LIF) has been done, perhaps providing an expressed molecular milieu which could identify candidate stem cells. Attempts to determine the complete molecular signature of gene expression common to human and mouse stem cells have shown over 200 common genes between hematopoietic and neural stem cells, with some considerable overlap with mouse embryonic stem cells as well. The function of many of these genes is as yet unknown, but may provide distinctive markers for identification of adult stem cells in different tissues.

However, dependence on particular markers for prospective identification and isolation of adult stem cells seems unreliable. In particular, the use of specific hematopoietic markers such as the presence or absence of CD34, has yielded mixed results in terms of the identification of putative stem cells. There is evidence that the expression of CD34 and CD133 can actually change over time, and its expression may be part of a cycling phenomenon among human hematopoietic and mesenchymal stem cells in the bone marrow and peripheral blood, and perhaps in other tissues, *i.e.*, an isolated CD34 cell may become CD34 , and then reacquire CD34 expression.

Likewise, a systematic analysis of the cell surface markers and differentiation potential of supposedly distinct isolated populations of human bone marrow stem cells revealed no differences in practice between the cell populations. Moreover, an analysis of genetic and ultrastructural characteristics of human mesenchymal stem cells undergoing differentiation and dedifferentiation has revealed reversibility in the characteristics studied. Thus, any attempt to isolate a single type of adult stem cell for study may not actually capture the intended cell, or may, by using a particular set of isolation or growth conditions, alter its gene expression. This idea has been elaborated by Thiese and Krause, who note that this "uncertainty principle" means any attempt to isolate and characterize a cell necessarily alters its environment, and thereby potentially its gene expression, identity, and potential ability to differentiate along various lineages. Likewise, the stochastic nature of cell differentiation in such dynamic and interacting systems means that attempts to delineate differentiation pathways must include descriptions of each parameter associated with the conditions used, and still may lead only to a probabilistic outcome for differentiation of a stem cell into a particular tissue. Blau *et al.* have raised the question of whether there may be a "universal" adult stem cell, residing in multiple tissues and activated dependent on cellular signals, *e.g.,* tissue injury. When recruited to a tissue, the stem cell would take its cues from the local tissue milieu in which it finds itself (including the soluble growth factors, extracellular matrix, and cell-cell contacts.) Examples of such environmental influences on fate choice have been noted previously. Thus, it may not be surprising to see examples of cells isolated using the same marker set showing disparate differentiative potentials, based on the context of the isolation or experimental conditions, or to see cells with different marker sets showing similar differentiation. In the final analysis, description of a "stem cell", its actual tissue of origin, and even its differentiation ability, may be a moving target describable only within the context of the particular experimental paradigm used, and may require asking the correct questions in context of the cell's identity and abilities not clonally but rather within a population of cells, and within a certain environment. Given the uncertainties involved in isolating and identifying particular adult stem cells, Moore and Quesenberry suggest that we consider an adult stem cell's functional ability to be, at a minimum, taking on the morphology and cell markers of a differentiated tissue, supplemented by any further functional activity and interaction within a tissue. Certainly a physiological response by improvement of function in a damaged organ system is an indication of a functional response. As will be discussed later, the function and therapeutic benefit may not necessarily require direct differentiation and integration of an adult stem cell into a desired tissue, but could be accomplished by stimulation of endogenous cells within the tissue.

Differentiation Mechanisms

Several possible mechanisms have been proposed for differentiation of adult stem cells into other tissues. One mechanism that has received attention lately is the possibility of cell fusion, whereby the stem cell fuses with a tissue cell and takes on that tissue's characteristics. *In vitro* experiments using fusion of somatic cells with embryonic stem cells and embryonic germ cells have demonstrated that the cell hybrid can take on characteristics of the more primitively developed cell. However, given that such characteristics of spontaneous cell fusion hybrids *in vitro* have been known for quite some time, and that a cell fusion hybrid does not explain *in vitro* differentiation of adult stem cells unexposed to tissues, the experiments could not verify this as a possible mechanism for adult stem cell differentiation. More recently, *in vivo* experiments have shown that for liver, formation of a cell fusion hybrid is a viable explanation for some of the differentiation as well as repair of liver damage seen in these experiments. In an *in vitro* experiment where human mesenchymal stem cells were co-cultured with heat-shocked small airway epithelial cells, a mixed answer was obtained—some of the stem cells differentiated directly into epithelial cells, while others formed cell fusion hybrids to repair the damage. The ability to form cell hybrids in some tissues may be a useful mechanism for repair of certain types of tissue damage or for delivery of therapeutic genes to a tissue. The reprogramming of cellular gene expression via hybrids is not unlike a novel method reported recently for transdifferentiation of somatic cells. In this method, fibroblasts were soaked in the cytoplasm and nucleoplasm of a lysed, differentiated T lymphocyte cell, taking up factors from the exposed "soup" of the cellular contents of the differentiated cell, and began expressing functional characteristics of a T cell. In contrast to the results discussed above, other experiments have shown no evidence that cell fusion plays a role in differentiation of adult stem cells into other tissue types. For example, using human subjects it was shown that human bone marrow cells differentiated into buccal epithelial cells *in vivo* without cell fusion, and human cord blood stem cells formed hepatocytes in mouse liver without evidence of cell fusion. In these cases it appears that the adult stem cells underwent changes in gene expression and directly differentiated into the host tissue cell type, integrating into the tissue. It is likely that the mechanism of adult stem cell differentiation may vary depending on the target tissue, or possibly on the state of the adult stem cell used, especially given that normal functioning liver typically shows cell fusion hybrids, with cell fusion functioning as a mechanism for most of the differentiation and repair in tissues such as liver, and direct differentiation (transdifferentiation) into other cell types functioning in other tissues. Much remains to be determined regarding the mechanisms associated with adult stem cell differentiation.

Keeping in mind the uncertainties noted above for identification of a particular adult stem cell and its initial tissue of origin, the majority of this review will focus on some of the evidence for adult stem cell differentiation into other tissues. The cells will be categorized based on general tissue of isolation, with the primary emphasis on human adult stem cells, supplemented with information from animal studies.

Bone Marrow Stem Cells

Bone marrow contains at least two, and likely more, discernable stem cell populations. Besides the hematopoietic stem cell which produces blood cell progeny, a cell type termed mesenchymal or stromal also exists in marrow. This cell provides support for hematopoietic and other cells within the marrow, and has also been a focus for possible tissue repair. Isolation is typically based on some cell surface markers, but also primarily on the ability of these cells to form adherent cell layers in culture. Human mesenchymal stem cells have been shown to differentiate *in vitro* into various cell lineages including neuronal cells, as well as cartilage, bone, and fat lineages. *In vivo*, human adult mesenchymal stem cells transferred *in utero* into fetal sheep can integrate into multiple tissues, persisting for over a year. The cells differentiated into cardiac and skeletal muscle, bone marrow stromal cells, fat cells, thymic epithelial cells, and cartilage cells. Analysis of a highly purified preparation of human mesenchymal stem cells indicated that they could proliferate extensively in culture, constitutively expressing the telomerase enzyme, and even after extensive culture retained the ability to differentiate *in vitro* into bone, fat, and cartilage cells. Isolated colonies of the cells formed bone when injected into immunodeficient mice. Expanding on their previous *in vitro* work with rat and human mesenchymal/stromal stem cells, Woodbury *et al.* performed molecular analyses of rat stromal stem cells and found that the cells express genes associated with all three primary germ layers—mesodermal, ectodermal, and endodermal—as well as a gene associated for germinal cells. The gene expression pattern was also seen in a clonal population of cells, indicating that it was not due to an initial mixed population of cells, but was the typical gene expression pattern of the stromal cells. The results suggested that the stromal stem cells were already multidifferentiated and that switching to a neuronal differentiation pattern involved quantitative regulation of existing gene expression patterns. Koc *et al.* have used infusion of allogeneic donor mesenchymal stem cells in an attempt to correct some of the skeletal and neurological defects associated with Hurler syndrome (mucopolysaccharidosis type-IH) and metachromatic leukodystrophy (MLD). A total of 11 patients received donor mesenchymal stem cells, expanded from bone marrow aspirate. Four patients showed significant improvements in nerve conduction velocities, and all patients showed maintenance or slight improvement in bone mineral density.

Bone marrow-derived cells in general have shown ability to form many tissues in the body. For example, bone marrow-derived stem cells *in vivo* appear able to form neuronal tissues, and a single adult bone marrow stem cell can contribute to tissues as diverse as marrow, liver, skin, and digestive tract. One group has now developed a method for large-scale generation of neuronal precursors from whole adult rat bone marrow. In this procedure, treatment of unfractionated bone marrow in culture with epidermal growth factor and basic fibroblast growth factor gave rise to neurospheres with cells expressing neuronal markers.

In vivo studies using fluorescence and genetic tracking of adult stem cells in animals, and tracking of the Y chromosome in humans, has shown that bone marrow stem cells can contribute to numerous adult tissues. Follow-up of patients receiving adult bone marrow stem cell transplants has allowed tracking of adult stem cells within humans, primarily by identification of Y chromosomebearing cells in female patients who had received bone marrow stem cells from male donors. Biopsy or postmortem samples show that some of the transplanted bone marrow stem cells could form liver, skin, and digestive tract cells, as well as participate in the generation of new neurons within the human brain. Bone marrow stem cells have also been shown to contribute to Purkinje cells in the brains of adult mice and humans . Generation of this particular type of neural cell is significant in that new Purkinje cells do not normally appear to be generated after birth.

Regeneration or replacement of dead or damaged cells is the primary goal of regenerative medicine and one of the prime motivations for study of stem cells. It is thus of significant interest that bone marrow stem cells have shown the ability to produce therapeutic benefit in animal models of stroke. In mice, fluorescencetracked bone marrow derived stem cells expressed neuronal antigens and also incorporated as endothelial cells, possibly producing therapeutic benefit by allowing increased blood flow to damaged areas of the brain. In rats, intravenous (IV) administration of rat or human bone marrow stromal cells resulted in significant behavioral recovery after stroke. Interestingly, only a small percentage of the stromal stem cells appeared to incorporate into the damaged brain as neuronal cells (1-5% in the case of the human marrow stromal cells), but the levels of neurotrophin growth factors within the brains increased and were possibly the signal for repair of damaged brain tissue, perhaps by stimulation of endogenous neuronal precursors. It is also of interest that the marrow stromal cells were injected IV and not intracerebrally, indicating that the stem cells somehow "homed" to the site of tissue damage. Most studies showing adult stem cell differentiation into other tissues show an increased incorporation of cells, or even an absolute requirement for differentiation, relying on tissue damage to initiate the differentiation. This may indicate that without a "need" for replacement and repair, there is little or no activation of adult stem cells. The recruitment and homing of adult stem cells to damaged tissues are fascinating but relatively unexplained phenomena. One report indicates that recruitment of quiescent stem cells from bone marrow to the circulation requires release of soluble c-kit ligand (stem cell factor), but the range of factors necessary for recruitment and homing to organs other than bone marrow is unknown at this time and warrants increased investigation.

Bone marrow stem cells have also shown the ability to participate in repair of damaged retinal tissues. When bone marrow stem cells were injected into the eyes of mice, they associated with retinal astrocytes and extensively incorporated into the vascular (blood vessel) network of the eye. The cells could also rescue and maintain normal vasculature in the eyes of mice with a degenerative vascular disease. In another animal study, bone marrow derived stem cells were observed to integrate into injured retina and differentiated into retinal neuronal cells. Stromal stem cells have also shown capability in mice to repair spinal cord which was demyelinated. One of the problems related to spinal cord injury is loss of the protective myelin sheath from spinal cord after injury. A mixed bone marrow stem cell fraction was injected into the area of damage in the spinal cord, and remyelination of the area was seen.

In another mouse study, marrow stromal cells injected into injured spinal cord formed guiding strands within the cord; interestingly, the effect was more pronounced when the stromal cells were injected 1 week after injury rather than immediately after injury.

Because bone marrow stem cells are of mesodermal lineage, it is not surprising that they show capabilities at forming other tissues of mesodermal origin. Human marrow stromal cells, which have been shown to form cartilage cells, have been used in an *in vitro* system to define many of the molecular events associated with formation of cartilage tissue. Bone marrow derived stem cells have also been shown capable of regenerating damaged muscle tissue. In an elegant study following genetically marked bone marrow stem cells in mice, LaBarge and Blau were able to document multiple steps in the progression of the stem cells to form muscle fibers and repair muscle damage. The ability of human bone marrow derived stem cells to form muscle cells and persist in the muscle was recently documented. In this case, a patient had received a bone marrow transplant at age 1, and developed Duchenne muscular dystrophy at age 12. Biopsies at age 14 showed donor nuclei integrated within 0.5-0.9% of the muscle fibers of the patient, indicating the ability of donated marrow cells to persist in tissue over long periods of time. Bone marrow stem cells have also shown capability at forming kidney cells. Studies following genetically marked bone marrow stem cells in rats and mice showed that the stem cells could form mesangial cells to repopulate the glomerulus of the kidney. In the mouse study, formation of cell fusion products was ruled out as a mechanism for differentiation of the bone marrow stem cells. Other animal studies have shown contribution of bone marrow stem cells to repair of damaged renal tubules in the kidney; taken together, animal studies indicate that bone marrow stem cells can participate in restoring damaged kidney tissue. Liver was one of the earliest tissues recognized as showing potential contribution to differentiated cells by bone marrow stem cells. Bone marrow stem cells have been induced to form hepatocytes in culture and liver-specific gene expression has been induced *in vitro* in human bone marrow stem cells. *In vivo*, bone marrow stem cells were able to incorporate into liver as hepatocytes and rescue mice from a liver enzyme deficiency, restoring normal liver function. Bone marrow stem cells also repopulated liver after irradiation of mice to destroy their bone marrow. Examination of livers of female patients who had received male bone marrow transplants, and male patients who had received female liver transplants, showed that similar repopulation of liver from bone marrow stem cells could take place in humans. Examination of the kinetics of liver repopulation by bone marrow stem cells in a mouse model indicated that the replacement was slow, with only small numbers of cells replaced by the bone marrow stem cells. As noted previously, two recent studies have found that replenishment of liver by bone marrow stem cells occurs primarily via cell fusion hybrid formation, even in repair of liver damage. A side-population of stem cells has been identified in mouse liver, similar to that seen in bone marrow.

This hepatic side-population, which contributes to liver regeneration, can be replenished by sidepopulation bone marrow stem cells. Pancreas and liver arise from adjacent endoderm during embryological development, and show relatedness in some gene expression and interconversion in some instances. Bone marrow derived cells have shown the ability to form pancreatic cells in animal studies. Mouse bone marrow stem cells containing a genetic fluorescent marker that is only expressed if insulin is expressed were transplanted into irradiated female mice. Within 6 weeks of transplant, fluorescent donor cells were observed in pancreatic islets; donor cells identified in bone marrow and peripheral blood did not show fluorescence. *In vitro*, the bone marrow derived cells showed glucose-dependent insulin secretion as well. Bone marrow derived stem cells have also demonstrated the ability to induce regeneration of damaged pancreas in the mouse. Mice with experimentally induced hyperglycemia from pancreatic damage were treated with bone marrow derived stem cells expressing the c-kit marker. Interestingly, only a low percentage of donor cells were identified as integrating into the regenerating pancreas, with most of the regeneration due to induced proliferation and differentiation of endogenous pancreatic cell precursors, suggesting that the bone marrow stem cells provided growth signals for the tissue regeneration. Heart, as a mesodermally-derived organ, is a likely candidate for regeneration with bone marrow derived stem cells. Numerous references now document the ability of these adult stem cells to contribute to regeneration of cardiac tissue and improve performance of damaged hearts. In animal studies, for example, rat , mouse and human stem cells have been identified as integrating into cardiac tissue, forming cardiomyocytes and/or cardiac blood vessels, regenerating infarcted heart tissue, and improving cardiac function.

In mice, bone marrow derived stem cells injected into old animals seems capable of restoring cardiac function, apparently through increased activity for cardiac blood vessel formation. One fascinating study using xenogeneic (cross-species) transplants suggests that stromal cells may show immune tolerance by the host. Mouse marrow stromal cells were transplanted into fully immunocompetent rats, and contributed formation of cardiomyocytes and cardiac vessels. Even after 13 weeks, the mouse cells were not rejected by the rat hosts. Evidence has accumulated from postmortem studies that bone marrow stem cells can contribute to cardiomyocytes after damage to the human heart as well. The evidence has led numerous groups to use bone marrow derived stem cells in treatment of patients with damaged cardiac tissue. Results from these clinical trials indicate that bone marrow derived stem cells, including cells from the patients themselves, can regenerate damaged cardiac tissue and improve cardiac performance in humans. In terms of restoring angiogenesis and improving blood circulation, results in patients are not limited to the heart. Tateishi-Yuyama *et al.* have shown that bone marrow derived stem cells from the patients themselves can improve blood circulation in gangrenous limbs, in many cases obviating the need for amputation.

Bone marrow derived adult stem cells have also been found to contribute to various other adult tissues. Animal studies indicate evidence that bone marrow stem cells can contribute as progenitors of lung epithelial tissue , and mesenchymal stem cells can home to damaged lung tissue, engraft, and take on an epithelial morphology, participating in repair and reduction of inflammation. Bone marrow derived stem cells also have been shown to contribute to regeneration of gastrointestinal epithelia in human patients. A recent study in mice has indicated that bone marrow stem cells can also participate in cutaneous healing, contributing to repair of skin after wounding.

Periheral Blood Stem Cells

There is abundant evidence that bone marrow stem cells can leave the marrow and enter the circulation, and specific mobilization of bone marrow stem cells is used to harvest stem cells more easily for various bone marrow stem cell treatments. Therefore, it is not surprising that adult stem cells have been isolated from peripheral blood. Mobilized stem cells in peripheral blood have been administered intravenously in a rat model of stroke, ameliorating some of the behavioral deficits associated with the damaged neural tissue , leading to a proposal that stem cell mobilization in patients might be used as a treatment for stroke in humans. Mobilized stem cells have also been used in cardiac regeneration in mice . Two recent studies have found that human peripheral blood stem cells exhibiting pluripotent properties can be isolated from unmobilized human blood. One study showed that the isolated cells were adherent, similar to marrow mesenchymal cells, and could be induced to differentiate into cells from all three primary germ layers, including macrophages, T lymphocytes, epithelial cells, neuronal cells, and liver cells. The other study showed induction of the peripheral blood stem cells could produce hematopoietic, neuronal, or cardiac cells in culture. In the latter study, undifferentiated stem cells were negative for both major histocompatability antigens (MHC) I and II, expressed high levels of the Oct-4 gene (usually associated with pluripotent capacity in other stem cells), and could form embryoid body structures in culture.

Neuronal Stem Cells

One extremely interesting finding of the past few years has been the discovery of neuronal stem cells, indicating that cell replenishment was possible within the brain (something previously considered impossible.) Neuronal stem cells have been isolated from various regions of the brain including the more-accessible olfactory bulb as well as the spinal cord , and can even be recovered from cadavers soon after death. Evidence now exists that neuronal stem cells can produce not only neuronal cells but also other tissues, including blood and muscle. Animal studies have shown that adult neural stem cells can participate in repair of damage after stroke, either via endogenous neuronal precursors or transplanted neural stem cells. Evidence indicates that endogenous neurons and astrocytes may also secrete growth factors to induce differentiation of endogenous precursors. In addition, two studies now provide suggestive evidence that neural stem cells/neural progenitor cells may show low immunogenicity, being immunoprivileged on transplant, and raising the possibility for use of donor neural stem cells to treat degenerative brain conditions.

Pluchino *et al.* recently used adult neural stem cells to test potential treatment of multiple sclerosis lesions in the brain.

Using a mouse model of chronic multiple sclerosis—experimental immune encephalitis—they injected neural stem cells either intravenously or intracerebrally into affected mice. Donor cells entered damaged, demyelinated regions of the brain and differentiated into neuronal cells. Remyelination of brain lesions and recovery from functional impairment were seen in the mice.

Neural stem cells have also been used to investigate potential treatments for Parkinson's disease. Using experimentally-lesioned animals as models for Parkinson's disease, human neural stem cells have been observed to integrate and survive for extended periods of time.

Dopaminergic cells (the cells degenerated in Parkinson's disease) can be induced in these systems, and neural stem cells are capable of rescuing and preventing the degeneration of endogenous dopaminergic neurons, also producing improved behavioral performance in the animals. In these studies, the data suggest that the transplanted neural stem cells did not participate to a large extent in direct formation of dopaminergic neurons, but rather secreted neuroprotective factors and growth factors that stimulated the endogenous neural cells. In this respect, infusion of transforming growth factor into the brains of Parkinson's mice induced proliferation and differentiation of endogenous neuronal precursors in mouse brain. Following this potential for stimulation of endogenous neuronal cells, Gill *et al.* recently reported on a Phase I trial in which glial derived neurotrophic factor (GDNF) was infused into the brains of five Parkinson's patients. After one year there was a 61% increase in the activities of daily living score, and an increase in dopamine storage observed in the brain. In a tantalizing clinical application with direct injection of neural stem cells, a Parkinson's patient was implanted with his own neural stem cells, resulting in an 80% reduction in symptoms at one year after treatment. Further clinical trials are underway.

The olfactory ensheathing glial (OEG) cell from olfactory bulb has been used extensively in studies regarding spinal cord injury and axon regrowth. Human OEG cells can be expanded in number in culture and induced to produce all three main neural cell types. Transplant of the cells into animal models of spinal cord injury has shown that the cells can effect remyelination of demyelinated spinal cord axons, and provide functional recovery in paraplegic rats, including in transected spinal cords. Another study has found that infusion of growth factors such as GDNF can stimulate functional regeneration of sensory axons in adult rat spinal cord. Interestingly, one group has made use of the similarities between enteric glial cells and OEG cells, and shown that transplanted enteric glial cells can also promote regeneration of axons in the spinal cord of adult rats. Clinical trials are underway to test the abilities of OEG cells in spinal cord injury patients. Finally, a significant impediment to recovery from spinal cord injury is the formation of a glial/astrocyte scar at the site of injury, which can prevent growth of axons no matter what the source of the cells. Menet *et al.* have shown, using a mutant mouse model, that much of the scar can be prevented by inhibition of glial fibrillary acidic protein and vimentin. In mutant mice that lacked these genes, there was increased sprouting of axons and functional recovery after spinal cord injury. Thus, endogenous neural cell growth and reconnection might suffice for repair of damage if inhibitory mechanisms can be removed from neural systems.

hNT Cells

Embryonal carcinoma (EC) cells can be derived from teratocarcinomas of adult patients, and show multipotent differentiation abilities in culture. From one such isolation, a "tamed" (non-tumorigenic) line of cells with neuronal generating capacity has been developed, termed hNT (NT-2) cells. Because of their capacity to generate neuronal cells, these cells have been studied for possible application in regeneration of neuronal tissues. The hNT neurons show the ability to generate dopaminergic neurons, and have shown some benefit of transplantation in animal models of amyotrophic lateral sclerosis (ALS, Lou Gehrig's disease). Early clinical trials using hNT neurons transplanted into stroke patients have shown initial positive results.

Muscle Stem Cells

Muscle contains satellite cells that normally participate

in replacement of myoblasts and myofibers. There are also indications that muscle additionally may harbor other stem cells, either as hematopoietic migrants from bone marrow and peripheral blood, or as intrinsic stem cells of muscle tissue. Muscle appears to contain a side population of stem cells, as seen in bone marrow and liver, with the ability to regenerate muscle tissue. Muscle derived stem cells have been clonally isolated and used to enhance muscle and bone regeneration in animals. An isolated population of musclederived stem cells has also been shown to participate in muscle regeneration in a mouse model of muscular dystrophy. Stimulation of muscle regeneration from muscle-derived stem cells, as observed in other tissues, is greatly increased after injury of the tissue. An interesting use of muscle-derived stem cells has been the regeneration and strengthening of bladder in a rat model of incontinence. Because of the similar nature of muscle cells between skeletal muscle and heart muscle, muscle-derived stem cells have also been proposed for use in repairing cardiac damage, with evidence that mechanical beating is necessary for full differentiation of skeletal muscle stem cells into cardiomyocytes. At least one group has used skeletal muscle cells for clinical application to repair cardiac damage in a patient, with positive results.

Liver Stem Cells

As noted before, there are similarities between liver and pancreas which could facilitate interconversion of cells between the two tissues. This concept has been demonstrated using genetic engineering to add a pancreatic development gene to liver cells, converting liver to pancreas. Rat liver stem cells have been converted *in vitro* into insulin-secreting pancreatic cells. When transplanted into immundeficient mice which are a model for diabetes, the converted liver stem cells were able to reverse hyperglycemia in the mice.

One other interesting observation regarding liver stem cells has been the possible formation of myocytes in the heart by liver stem cells. A clonal cell line derived from adult male rat liver and genetically tagged was injected into female rats, and marked, Y-chromosome bearing myocytes were identified in the host hearts after six weeks.

Pancreatic Stem Cells

Interconversion between pancreas and liver has also been demonstrated starting with pancreatic stem cells, in which mouse pancreatic cells repopulated the liver and corrected metabolic liver disease. For pancreas, however, the possibility of solutions to the scourge of diabetes has been a driving force in efforts to define a stem cell that could regulate insulin in a normative, glucosedependent fashion. The success of the Edmonton protocol, where cadaveric pancreatic islets are transplanted into patients, has provided a glimmer of hope, but more readily-available sources of insulin-secreting cells are needed. Fortunately, there seems to be no shortage of potential candidates that can form insulin-secreting cells. The pancreas itself appears to contain stem/progenitor cells that can regenerate islets *in vitro* and *in vivo*. Studies indicate that these pancreatic stem cells can functionally reverse insulindependent diabetes in mice. Similar pancreatic stem cells have been isolated from humans and shown to form insulin-secreting cells *in vitro*, the hormone glucagon-like peptide-1 appears to be an important inducing factor of pancreatic stem cell differentiation.

Interestingly, the same hormone could induce mouse intestinal epithelial cells to convert into insulin-producing cells *in vitro*, and the cells could reverse insulin-dependent diabetes when implanted into diabetic mice. Besides pancreatic and intestinal stem cells, other adult stem cell types showing the ability to secrete insulin and regenerate damaged pancreas include bone marrow and liver. Genetic engineering of rat liver cells to contain the pancreatic gene PDX-1 has also been used to generate insulin-secreting cells *in vitro*; the cells could also restore normal blood glucose levels when injected into mice with experimentally-induced diabetes.

Corneal Limbal Stem Cells

Corneal limbal stem cells have become commonly used for replacement of corneas, especially in cases where cadaveric donor corneas are insufficient. Limbal cells can be maintained and cell number expanded in culture, grown on amniotic membranes to form new corneas, and transplanted to patients with good success. A recent report indicates that human corneal stem cells can also display properties of functional neuronal cells in culture. Another report found that limbal epithelial cells or retinal cells transplanted into retina of rats could incorporate and integrate into damaged retina, but did not incorporate into normal retina.

Mammary Stem Cells

Reports have indicated that mammary stem cells also exist.

Isolated cells from mouse could be propagated *in vitro* and differentiated into all three mammary epithelial lineages. Clonallypropagated cells were induced in culture to generate complex threedimensional structures similar to that seen *in vivo*. Transcriptional profiling indicated that the mammary stem cells showed similar gene expression profiles to those of bone marrow stem cells. In that respect, there is a report that human and mouse mammary stem cells exist as a side population, as seen for bone marrow, liver, and muscle stem cells. When propagated in culture, the isolated mammary side population stem cells could form epithelial ductal structures.

Salivary Gland

A recent report indicates that stem cells can be isolated by limiting dilution from regenerating rat salivary gland and propagated *in vitro*. Under differing culture conditions, the cells express genes typical of liver or pancreas, and when injected into rats can integrate into liver tissue.

Skin

Multipotent adult stem cells have been isolated from the dermis and hair follicle of rodents. The cells play a role in maintenance of epidermal and hair follicle structures, can be propagated *in vitro,* and clonally isolated stem cells can be induced to form neurons, glia, smooth muscle, and adipocytes in culture.

Dermal hair follicle stem cells have also shown the ability to reform the hematopoietic system of myeloablated mice.

Tendon

A recent report notes the isolation of established stem celllike lines from mouse tendon. The cells exhibited a mesenchymal morphology, and expressed genes related to osteogenic, chondrogenic, and adipogenic potential, similar to that seen in bone marrow mesenchymal stem cells.

Synovial Membrane

Stem cells from human synovial membrane (knee joint) have been isolated which show multipotent abilities for differentiation, including evidence of myogenic potential. These stem cells were used in a mouse model of Duchenne muscular dystrophy to test their ability to repair damaged muscle. Stem cells injected into the bloodstream could engraft and incorporate into muscle, taking on a muscle phenotype, and with evidence of muscle repair.

Heart

Beltrami *et al.* analyzed the hearts of post-mortem patients who succumbed 4-12 days after heart attack, and found evidence of dividing myocytes in the human heart. While it is unclear from the study whether the cells were originally cardiomyocytes or were other stem cells which had homed to damaged heart tissue, such as bone marrow stem cells, the evidence indicated dividing cells within the heart.

Cartilage

Human cartilage biopsies placed into culture show apparent dedifferentiation into primitive chondrocytes with mesenchymal stem cell appearance. These chondrocytes have been used for transplants to repair articular cartilage damage, and in treatment of children with osteogenesis imperfecta.

Thymic Prognitors

Bennett *et al.* have reported the isolation of thymic epithelial progenitor cells. Ectopic grafting (under the kidney capsule) of the cells into mice allowed production of all thymic epithelial cell types, as well as attraction of homing T lymphocytes. In separate experiments, Gill *et al.* also isolated a putative thymic progenitor cell from mice and were able to use these cells to reform miniature thymuses when the cells were transplanted under mouse kidney capsule.

Dental Pulp Stem Cells

Stem cells have been isolated from human adult dental pulp that could be clonally propagated and proliferated rapidly. Though there were some similarities with bone marrow mesenchymal stem cells, when injected into immunodeficient mice the adult dental pulp stem cells formed primarily dentin-like structures surrounded by pulpy interstitial tissue. Human baby teeth have also been identified as a source of stem cells, designated SHED cells (Stem cells from Human Exfoliated Deciduous teeth). *In vitro*, SHED cells could generate neuronal cells, adipocytes, and odontoblasts, and after injection into immunodeficient mice, the cells were indicated in formation of bone, dentin, and neural cells.

Adipose (FAT) Derived Stem Cells

One of the more interesting sources identified for human stem cells has been adipose (fat) tissue, in particular liposuctioned fat. While there is some debate as to whether the cells originate in the fat tissue or are perhaps mesenchymal or peripheral blood stem cells passing through the fat tissue, they represent a readily-available source for isolation of potentially useful stem cells. The cells can be maintained for extended periods of time in culture, have a mesenchymal-like morphology, and can be induced *in vitro* to form adipose, cartilage, muscle, and bone tissue. The cells have also shown the capability of differentiation into neuronal cells.

Umbilical Cord Blood

Use of umbilical cord stem cells has seen increasing interest, as the cells have been recognized as a useful source for hematopoietic transplants similar to bone marrow stem cell transplants, including for treatment of sickle cell anemia. Cord blood shows decreased graft-versus-host reaction compared to bone marrow, perhaps due to high interleukin-10 levels produced by the cells. Another possibility for the decreased rejection seen with cord blood stem cell transplants is decreased expression of the beta-2-microglobulin on human cord blood stem cells. Cord blood can be cryopreserved for over 15 years and retain significant functional potency. Cord blood stem cells also show similarities with bone marrow stem cells in terms of their potential to differentiate into other tissue types. Human cord blood stem cells have shown expression of neural markers *in vitro*, and intravenous administration of cord blood to animal models of stroke has produced functional recovery in the animals. Infusion of human cord blood stem cells has also produced therapeutic benefit in rats with spinal cord injury, and in a mouse model of ALS. A recent report noted establishment of a neural stem/progenitor cell line derived from human cord blood that has been maintained in culture over two years without loss of differentiation ability. Several reports also note the production of functional liver cells from human cord blood stem cells. Additional differentiative properties of human umbilical cord blood stem cells are likely to be discovered as more investigation proceeds on this source of stem cells.

Umbilical Cord Mesenchyme (Wharton's Jelly)

While most of the focus regarding umbilical cord stem cells has focused on the cord blood, there are also reports that the matrix cells from umbilical cord contain potentially useful stem cells. Using pigs, this matrix from umbilical cord, termed Wharton's jelly, has been a source for isolation of mesenchymal stem cells. The cells express typical stem cell markers such as c-kit and high telomerase activity, have been propagated in culture for over 80 population doublings, and can be induced to form neurons *in vitro*. When transplanted into rats, the cells expressed neuronal markers and integrated into the rat brain, additionally without any evidence of rejection.

Amniotic Stem Cells

Amniotic fluid has also been found to contain stem cells that can take on neuronal properties when injected into brain. These stem cells were recently isolated from human amniotic fluid, and were found to express Oct-4, a gene typically associated with expression in pluripotent stem cells.

Mesangioblasts

Mesangioblasts are a multipotent stem cell that has been isolated from large blood vessels such as dorsal aorta. The cells show long term proliferative capacity in culture as well as the capability of differentiation into most mesodermally derived types of tissue. In a recent report, the cells were injected into the bloodstream of mice that are a model for muscular dystrophy, and participated in repair of the muscle tissue.

Adult stem cells in other tissues very likely exist, but this survey of many of the known adult stem cells and their capacities for differentiation and tissue repair can serve as a beginning point for discussion regarding the progress as well as potential of adult stem cells. Some final thoughts on current and potential utilization of adult stem cells follow.

Adult Stem Cell Mobilization for Tissue Repair

An important point to consider as we look ahead regarding utilization of adult stem cells for tissue repair is that it may be unnecessary first to isolate and culture stem cells before injecting them back into a patient to initiate tissue repair. Rather, it may be easier and preferable to mobilize endogenous stem cells for repair of damaged tissue. Initial results regarding this possibility have already been seen in some animal experiments, in which bone marrow and peripheral blood stem cells were mobilized with injections of growth factors and participated in repair of heart and stroke damage. The ability to mobilize endogenous stem cells, coupled with natural or perhaps induced targeted homing of the cells to damaged tissue, could greatly facilitate use of adult stem cells in simplified tissue regeneration schemes.

Gene Theraphy Applications with Adult Stem Cells

Adult stem cells can provide an efficient vehicle for gene therapy applications, and engineered adult stem cells may allow increased functionality, proliferative capacity, or stimulatory capability to these cells. The feasibility of genetically engineering adult stem cells has been shown, for example, in the use of bone marrow stem cells containing stably inserted genes. The engineered stem cells when injected into mice could still participate in formation and repair of differentiated tissue, such as in lung. As another example, engineered stem cells containing an autoantigen, to induce immune tolerance of T cells to insulin-secreting cells, were shown to prevent onset of diabetes in a mouse model of diabetes, a strategy that may be useful for various human autoimmune diseases.

Introduction of the PDX-1 gene into liver stem cells stimulated differentiation into insulin-producing cells which could normalize glucose levels when transplanted into mice with induced diabetes. Simply engineering cells to increase their proliferative capacity can have a significant effect on their utility for tissue engineering and repair. For example, McKee *et al.* engineered human smooth muscle cells by introducing human telomerase, which greatly increased their proliferative capacity beyond the normal lifespan of smooth muscle cells in culture, while allowing retention of their normal smooth muscle characteristics. These engineered smooth muscle cells were seeded onto biopolymer scaffolds and allowed to grow into smooth muscle layers, then seeded with human umbilical vein endothelial cells. The resulting engineered arterial vessels could be useful for transplants and bypass surgery. Similarly, human marrow stromal cells that were engineered with telomerase increased their proliferative capacity significantly, but also showed enhanced ability at stimulating bone formation in experimental animals. Genetically-engineered human adult stem cells have already been used in successful treatment of patients with genetic disease. Bone marrow stem cells, from infants with forms of severe combined immunodeficiency syndrome (SCID), were removed from the patients, a functional gene inserted, and the engineered cells reintroduced to the same patients. The stem cells homed to the bone marrow, engrafted, and corrected the defect. Adult stem cells could also be used to deliver stimulatory or protective factors to tissues and endogenous stem cells. This would utilize the innate homing ability of adult stem cells, but would not necessarily rely on differentiation of the stem cells to participate in tissue replenishment. For example, Benedetti *et al.* utilized the homing capacity of neural stem cells in brain by engineering mouse neural stem cells with the gene for interleukin-4. Transfer into brain glioblastomas in mice led to the survival of most of the mice, and imaging analysis documented the progressive disappearance of large tumors. Likewise, engineered mesenchymal stem cells were transplanted into the brains of mice that are a model of Niemann- Pick disease; the enzyme acid sphingomyelinase is lost in the disease, resulting in neurological damage and early death. The mesenchymal stem cells were engineered to overexpress the missing enzyme. When injected into brains of the mouse model, the mice showed a delay in onset of neurological abnormalities and an extension of lifespan, suggesting that the stem cells delivered and secreted the necessary enzyme to the brain tissue. Musclederived stem cells that were engineered to express the growth factor bone morphogenetic protein-2 were used to stimulate bone healing in mice with skull bone defects. While the muscle-derived stem cells did show differentiation as bone cells, the results indicated that the critical factor was delivery of the secreted growth factor by the stem cells to the areas of bone damage, allowing much more rapid healing than in control animals. As noted previously, neural stem cells show an ability to rescue degenerating neurons, including the dopaminergic neurons whose loss is associated with Parkinson's disease. The delivery of neuroprotective substances is postulated as the most likely explanation for this phenomenon, rather than substantial differentiation by the injected neural stem cells. In support of this hypothesis, when neural stem cells were specifically engineered to overexpress a neurotrophic factor similar to glial derived neurotrophic factor, degeneration of dopaminergic neurons was prevented.

Stimulating Endogenous Cells

The indications from the previous examples suggest that direct stimulation of endogenous stem cells within a tissue may be the easiest, safest, and most efficient way to stimulate tissue regeneration. Such stimulation need not rely on any added stem cells. This approach would circumvent the need to isolate or grow stem cells in culture, or inject any stem cells into the body, whether the cells were derived from the patient or another source. Moreover, direct stimulation of endogenous tissue stem cells with specific growth factors might even preclude any need to mobilize stem cells to a site of tissue damage. A few experimental results suggest that this approach might be possible. One group has reported that use of glial derived neurotrophic factor and neurotrophin-3 can stimulate regeneration of sensory axons in adult rat spinal cord.

Administration of transforming growth factor to the brains of Parkinson's mice stimulated proliferation and differentiation of endogenous neuronal stem cells and produced therapeutic results in the mice, and infusion of glial derived neurotrophic factor into the brains of Parkinson's patients resulted in increased dopamine production within the brain and therapeutic benefit to the patients. And, Zeisberg *et al.* have found that bone morphogenetic protein-7 (BMP-7) can counteract deleterious cell changes associated with tissue damage. In this latter study, a mouse model of chronic kidney damage was used. Damage to the tissue causes a transition from epithelial to mesenchymal cell types in the kidney, leading to fibrosis.

The transition appears to be initiated by the action of transforming growth factor beta-1 on the tissues, and BMP-7 was shown to counteract this signaling *in vitro*. Systemic administration of BMP- 7 in the mouse model reversed the transition *in vivo* and led to repair of severely damaged renal tubule epithelial cells. These experiments indicate that direct stimulation of tissues by the correct growth factors could be sufficient to prevent or repair tissue damage.

The key to such treatments would be identification of the correct stimuli specific to a tissue or cell type.

In summary, our current knowledge regarding adult stem cells has expanded greatly over what was known just a few short years ago. Results from both animal studies and early human clinical trials indicate that they have significant capabilities for growth, repair, and regeneration of damaged cells and tissues in the body, akin to a built-in repair kit or maintenance crew that only needs activation and stimulation to accomplish repair of damage. The potential of adult stem cells to impact medicine in this respect is enormous.

Adult Stem Cells—Addendum (October 2003) For the President's Council on Bioethics David A. Prentice Since initial submission of the commissioned paper, numerous additional published references have documented the abilities of adult stem cells to stimulate regeneration of damaged tissues. Just a few of the most significant are mentioned here. Mesenchymal stem cells engineered to express the *Akt1* gene, when transplanted into mice, demonstrated the ability to repair and restore performance of infarcted heart, essentially to a normal state. Another clinical trial in addition to those mentioned in the paper has shown significant improvement in patients with heart damage, with reduction in the area of damage and improved heart function after adult stem cell treatment. Three more published articles support the existence of a stem cell in the heart and its participation in cardiac regeneration. Stroke damage in rats was repaired using human neural stem cells and prostate was regenerated *in vivo* in mice using adult stem cells. Another report indicates that human mixed bone marrow stem cells can contribute significant amounts of lung tissue in patients and pluripotent stem cells were discovered in the mouse inner ear , which can form all 3 primary germ layers and might lead to potential therapies for hearing loss. Finally, bone marrow stem cells were discovered to have a protective as well as regenerative role in diabetes.

APPENDIX 6

STEM CELLS AND TISSUE REGENERATION: LESSONS FROM RECIPIENTS OF SOLID ORGAN TRANSPLANTATION

1. Overview

The Major Histocompatibility Complex (MHC) is located on the short arm of chromosome 6 in humans and encodes the alloantigens known as Human Leukocyte Antigens (HLA), polymorphic cell surface molecules which enable the immune system to recognize both self and foreign antigens. The class II HLA molecules (HLADR, HLA-DP, and HLA-DQ) are usually found only on antigen- 347 presenting cells such as B lymphocytes, macrophages, and dendritic cells of lymphoid organs, and initiate the immune response to foreign proteins, including viruses, bacteria, and foreign HLA antigens on transplanted organs.

Following binding of foreign proteins, class II HLA on antigenpresenting cells activate CD4+ T cells, which in turn activate cytotoxic CD8+ T cells to recognize the same foreign antigen bound to HLA class I (HLA-A, HLA-B, and HLA-C, molecules found on the surface of all cells) and destroy the target. The actual recognition of foreign HLA transplantation antigens by T cells is referred to as allorecognition. Two distinct pathways of allorecognition have been described, direct and indirect. The direct pathway involves receptors on the host T cells that directly recognize intact HLA antigens on the cells of the transplanted organ. The indirect pathway requires an antigen-presenting cell that internalizes the foreign antigen and presents it via its own HLA class II molecule on the surface of an antigen-presenting cell to the CD4+ helper T cells.

Once recognition has taken place, an important cascade of events is initiated at the cellular level, culminating in intracellular release of ionized calcium from intracellular stores. The calcium binds with a regulatory protein called calmodulin, forming a complex that activates various phosphatases, particularly calcineurin. Calcineurin dephosphorylates an important cytoplasmic protein called nuclear factor of activated T cells (NFAT), resulting in its migration to the nucleus and induction of the production of various cytokines such as IL-2. These cytokines recruit other T cells to destroy the transplanted organ, ultimately resulting in rejection and loss of the graft.

Immunosuppressive regimens used to prevent allograft rejection are aimed at inhibiting the various arms of the immune response, typically require multiagent combinations, and need to be maintained for the duration of life. The currently used armamentarium confers significant side-effect risks, including infectious and neoplastic complications. Moreover, despite success at preventing early allograft rejection, long-term survival of transplanted organs remains difficult to achieve and novel methods to achieve long-term tolerance are being actively sought.

Stem cells obtained from embryonic or adult sources differ from other somatic cells in that they express very low levels of HLA molecules on their cell surfaces. This endows these cell types with the theoretical potential to escape the standard mechanisms of immune rejection discussed above. However, under conditions that enable cellular differentiation in vitro and in vivo each of these stem cell populations acquires high level expression of HLA molecules, suggesting that their long-term survival following transplantation in vivo may be limited by typical immune rejection phenomena.

Recent experimental data, however, provide striking counterintuitive examples that stem cells from both embryonic and adult sources may evade the recipient's immune system and result in long-term engraftment in the absence of immunosuppression despite acquisition of surface HLA molecule expression. These observations may have significant impact on the emerging field of regenerative medicine.

Immunobiology of Organ Transplantation

The Human Leukocyte Antigens (HLA)

Differences between individuals which enable immune recognition of non-self from self are principally due to the extreme polymorphism of genes in the Major Histocompatibility Complex (MHC) on chromosome 6 in man which encode the cell surface HLA molecules.

These molecules are cell surface glycoproteins whose

biologic function is to bind antigenic peptides (epitopes) derived from viruses, bacteria, or cancer cells, and present them to T cells for subsequent immune recognition. Each HLA gene includes a large number of alleles and the peptide binding specificity varies for each different HLA allele. The 1996 WHO HLA Nomenclature Committee report lists more than 500 different HLA class I and class II alleles.

Crystallographic x-ray studies have demonstrated that the hypervariable regions encoded by polymorphic regions in the alleles correspond to HLA binding pockets which engage specific "anchor" residues of peptide ligands. One HLA molecule will recognize a range of possible peptides, whereas another HLA molecule will recognize a different range of peptides. Consequently, no two individuals will have the same capability of stimulating an immune response, since they do not bind the same range of immunogenic peptides. It is estimated that >99% of all possible peptides derived from foreign antigens are ignored by any given HLA molecule. Since in the absence of HLA polymorphism a large number of immunogenic peptides would not be recognized, the extensive HLA polymorphism in the population reduces the chance that a given virus or bacterium would not be recognized by a sizable proportion of the population, reducing the likelihood of major epidemics or pandemics.

T Cell Recognition of Antigen Presented By HLA Molecules

Since HLA molecules regulate peptide display to and activation of the immune system, considerable effort has been devoted to understanding the molecular basis of peptide-HLA interactions.

These issues are important for defining the biology of T cell antigen recognition and the properties of a protein that make it immunogenic or non-immunogenic. Specific antigen recognition by T cells is dependent on recognition by the T cell receptor of a threedimensional complex on the surface of antigen-presenting cells (APC) comprised of the HLA molecule and its bound peptide. The peptides are produced by complex antigen processing machineries within the APC (i.e. proteolytic enzymes, peptide transporters and molecular chaperones) which generate a pre-selected peptide pool for association with the HLA molecules. The different types of T cells require different HLA molecules for antigen presentation, socalled "HLA restriction" phenomena. T cell receptors on CD8+ cytotoxic T cells (CTLs) bind peptides presented by HLA class I molecules, whereas CD4+ T helper cells (Th) recognize peptides bound to HLA class II molecules. Of the 8-13 amino acid residues of a bound peptide within a class I or II HLA molecule, only three to four amino acid side chains are accessible to the T cell receptor, and a similar number of amino acids are involved in binding to the HLA molecule.

Thymic Education of T Cells

T cells mature in the thymus to appropriately respond to foreign pathogens without inadvertently attacking the host. Under the influence of various thymic resident cells and factors they elaborate, maturing T cells fall into two categories: those that are able to discriminate between self and non-self and can appropriately respond to foreign pathogens without inadvertently attacking the host, and those which are unable to appropriately discriminate between self and non-self. Dendritic cells have been implicated in the deletion, or inhibition, of T cells reactive to self-antigens, particularly in the thymus during T cell development or in peripheral lymphoid organs. The process of self/non-self discrimination by the maturing T cells is dependent on thymic dendritic cell (DC) presentation of self-antigens in the context of self-HLA molecules.

When maturing thymic T cells are highly reactive with self-antigen/ HLA complexes, they are deleted so that potentially autoreactive T cells will not be released into the periphery. If a particular foreign antigen can be presented in such a way in the thymus as to fool the maturing T cells into believing that the antigen is part of self tissue, then T cells capable of reacting with this antigen will also be eliminated. Indeed, it has been demonstrated that when mouse thymic DC present transgenically introduced foreign antigens to developing T cells, the mature peripheral T cell repertoire of the mouse lacks T cells capable of reacting with the specific foreign antigen, i.e. it is tolerant to the foreign antigen. This has raised the possibility that injection of dendritic cells into an allogeneic recipient might induce tolerance to a subsequent allograft by causing deletion or inhibition of alloreactive T cells.

T Cell Recognition of Alloantigens

Recognition of foreign, or allogeneic, HLA antigens by the recipient immune system is the major limitation to the survival of solid organ grafts. The central role of HLA molecules in allograft rejection is due to their role as restriction elements for T cell recognition of donor antigens and the extensive polymorphism displayed by the HLA molecules, which elicit host immune responses. Although progress has been made in the short-term survival of transplants, chronic immunologic rejection remains an impediment to long-term survival.

The primary cause of acute rejection of transplanted organs is so-called "direct" recognition of whole allogeneic HLA antigens by receptors on the surface of recipient T cells. The direct recognition pathway involves recognition by recipient T cells of donor HLA class I and class II molecules, resulting in the generation of cytotoxic and helper T lymphocytes which play a pivotal role in the rejection process. In contrast, chronic rejection of transplanted organs results from so-called "indirect recognition" of donor HLA peptides derived from the allogeneic HLA molecules shed by the donor tissue. These foreign HLA molecules are taken up and processed by recipient antigen presenting cells (APC), and peptide fragments of the allogeneic HLA molecules containing polymorphic amino acid residues are bound and presented by recipient's (self) HLA molecules to recipient (self) T cells. Although direct and indirect recognition of alloantigen generally leads to adverse graft outcome, tolerance induction may occur following exposure of the recipient to donor alloantigens prior to transplantation. Since this strategy is based on the nature and dose of the antigen as well as the route of administration, understanding how to control the balance between activation and unresponsiveness mediated by the direct and/or indirect recognition of alloantigen is a an area of active research which could lead to development of new therapies to prolong graft survival.

Indirect allorecognition has been implicated in recurrent rejection episodes in various transplantation models of cardiac, kidney and skin grafts. Determinants on donor HLA molecules can be divided into two main categories: (a) the dominant allodeterminants that are efficiently processed and presented to alloreactive T cells during allograft rejection; and (b) the cryptic allodeterminants that are potentially immunogenic but do not normally induce alloreactive responses, presumably due to incomplete processing and/or presentation. Indirect recognition of allo-HLA peptides is important for the initiation and spreading of the immune response to other epitopes within the allograft. So-called "spreading" of indirect T cell responses to other allo-HLA epitopes expressed by graft tissue is strongly predictive of recurring episodes of rejection. Tolerance induction to the dominant donor determinants represents potential effective strategy for blocking indirect alloresponses and ensuring long-term graft survival in animal models.

Tolerance Induction

Advances in surgical methods and current immunosuppressive therapies have led to significant improvement in short-term graft survival, however long-term survival rates remain poor. For example, whereas both kidney and heart allografts have one-year graft survival rates of 85 to 95 percent, only about 50% of transplanted hearts survive five years and only about 50% of kidney grafts survive ten years. Thus, despite being able to achieve short-term success, these relatively poor long-term graft survival rates demonstrate the limitations of the current clinical immunosuppressive regimens to enable long-term immune evasion by the graft. Consequently, a major goal of transplantation immunobiologists is to induce donorspecific tolerance, allowing the long-term survival of human allografts without the need of HLA-compatibility and without the continuous recipient immunosupression leading to the concomitant risks of infection, malignancy, and/or other specific drug side effects.

This would theoretically improve long-term graft survival, reduce or eliminate the continuing need for expensive, toxic and non-specific immunosuppressive therapy and enhance the quality of life.

Insight into some of the mechanisms involved in tolerance induction has been gained from pre-clinical and clinical studies in numerous animal models and in patients, particularly those with liver allografts which typically do not induce a prominent immune response leading to rejection. One possible mechanism by which liver transplantation results in allograft tolerance tolerance may be that the donor or "passenger" lymphoid cells in the transplanted liver emigrate and take up residence in the recipient's immune organs, such as the thymus or lymph nodes. Donor lymphocytes at these sites might "re-educate" the recipient immune system so that the donor organ is not recognized as foreign. In an attempt to initiate a similar process in other organ recipients, transfusions of donor blood or bone marrow have been used to enhance solid organ graft survival in animal models and in clinical trials. These studies are currently ongoing in various organ systems.

Molecular understanding of the cellular immune response has led to new strategies to induce a state of permanent tolerance after transplantation. Several approaches have shown promise, including the use of tolerizing doses of class I HLA-molecules in various forms for the induction of specific unresponsiveness to alloantigens, and the use of synthetic peptides corresponding to HLA class II sequences. Other approaches include alteration in the balance of cytokines that direct the immune response away from the TH1 type of inflammatory response and graft rejection to the TH2 type of response that might lead to improved graft survival, and the use of agents to induce "co-stimulatory blockade" of T cell activation. This latter approach is based on the concept that blockade of a "second signal" to the T cell enables the signal provided to the T cell receptor by the HLA-peptide complex to induce antigen specific tolerance.

The experimental use of human dendritic cells as tolerogenic agents has been limited due to the low frequency of circulating dendritic cells in peripheral human blood, the limited accessibility to human lymphoid organs, and the terminal state of differentiation of circulating human dendritic cells making their further expansion ex vivo difficult. Dendritic cells are migratory cells of sparse, but widespread, distribution in both lymphoid and non-lymphoid tissues.

Although the earliest precursors are ultimately of bone marrow origin, the precise lineage of dendritic cells is controversial and includes both myeloid-derived and lymphoid-derived populations.

Recent work has revealed that an expanded population of mature human dendritic cells can be derived from non-proliferating precursors in vitro is by culturing bone-marrow derived cells with a combination of cytokines. This method of enrichment for human dendritic cells from a precursor population can result in the production of dendritic cells that are tolerogenic to foreign antigens.

Whether such cells could be useful when co-administered with an allograft transplant remains to be determined. Nevertheless, it is clear that considerable progress has been made in the past few years using approaches to manipulate the immune response to enable routine donor-specific tolerance, and there is reason to be optimistic that with better understanding of molecular and cellular mechanisms this goal could be attained.

Immonosuppressive Agends Commonly Used in Organ Transplant Recipients: Benefits and Adverse Outcomes

Cyclosporine

Cyclosporine has been the single most important factor associated with improved outcomes after organ transplantation over the past two decades. CyA binds to a cytosolic cell protein, cyclophilin (CyP).

The CyA-CyP complex then binds to calcineurin and subsequently blocks interleukin-2 (IL-2) transcription. The binding of IL-2 to the IL-2 receptors on the surface of T lymphocytes is a key stimulant in promoting lymphocyte proliferation, activation, and ultimately allograft rejection. A review of the first decade of experience with heart transplantation revealed a total of 379 cardiac allograft recipients worldwide; actuarial survival rates in this cohort of patients at 1 year and 5 years were 56% and 31% respectively; the main causes of death being acute rejection and the side effects of immunosuppression. With the introduction and widespread use of CyA over the next decade, survival rates dramatically improved to 85% and 75% at 1 and 5 years respectively. Similar results were obtained with other organ transplants, including kidney and lung.

The major adverse effects of CyA are nephrotoxicity, hypertension, neurotoxocity and hyperlipidemia; less common side effects include hirsuitism, gingival hyperplasia and liver dysfunction.

CyA nephrotoxicity can manifest as either acute or chronic renal dysfunction. It is important to note that a number of drugs commonly used in transplant patients, such as aminoglycosides, amphotericin B and ketoconazole can potentiate the nephrotoxicity induced by CyA. More than half the patients receiving CyA will require treatment for hypertension within the first year following transplantation.

Corticosteroids also potentiate the side effects of CyA such as hypertension, hyperlipidemia and hirsuitism. Frequent monitoring of the serum level is essential to minimize the adverse effects. One of the major limitations of the original oil-based CyA formulation (Sandimmune) is its variable and unpredictable bioavailability. In the mid-90s Neoral was introduced, a new microemulsion formula of CyA, which has greater bioavailability and more predictable pharmacokinetics than Sandimmune.

Tacrolimus

Tacrolimus (FK506) is a macrolide antibiotic that inhibits T-cell activation and proliferation and inhibits production of other cytokines. The product of *Streptomyces tsurubaensis* fermentation, FK 506 was first discovered in 1984 and first used in clinical studies in 1988 at the University of Pittsburgh. While the mechanism of action of tacrolimus is similar to that of CyA, and comparative clinical trials have suggested similar efficacy, it has been suggested that some groups of patients may benefit from tacrolimus rather than CyA as primary immunosuppressive therapy. Unlike CyA, hirsuitism and gingival hyperplasia occur infrequently with tacrolimus; thus, tacrolimus-based therapy may improve compliance and quality of life in female and pediatric transplant recipients. It should be noted that alopecia has been documented with tacrolimus, but is known to improve with dose reductions. The decreased incidence of hypertension and hyperlipidemia with tacrolimus makes it preferable to CyA in patients with difficult to treat hypertension or hyperlipidemia. A final indication for tacrolimus has been as a rescue immunosuppressant in cardiac transplant recipients on CyA with refractory rejection or intolerance to immunosuppression (severe side effects). Since tacrolimus is metabolized using the same cytochrome P450 enzyme system as CyA, drug interactions are essentially the same. Thus, drugs that induce this system may increase the metabolism of tacrolimus, thereby decreasing its blood levels.

Conversely, drugs that inhibit the P450 system decrease the metabolism of tacrolimus, thereby increasing its blood levels. It is important to note that some studies have indicated a higher incidence of nephrotoxicity with tacrolimus as compared to CyA.

Azathioprine and Mycophenolate Mofetil (MMF)

Despite being available for more than 35 years, azathioprine is still a useful agent as an immunosuppressive agent. Following administration, azathioprine is converted into 6-mercaptopurine, with subsequent transformation to a series of intracellularly active metabolites. These inhibit both an early step in *de novo* purine synthesis and several steps in the purine salvage pathway. The net effect is depletion of cellular purine stores, thus inhibiting DNA and RNA synthesis, the impact of which is most marked on actively dividing lymphocytes responding to antigenic stimulation. In currently used immunosuppressive protocols, azathioprine is used as part of a triple therapy regimen along with CyA or tacrolimus and prednisone. Mycophenolate mofetil (MMF), which is rapidly hydrolyzed after ingestion to mycophenolic acid, is a selective, noncompetitive, reversible inhibitor of onosine monophosphate dehydrogenase, a key enzyme in the de novo synthesis of guanine nucleotides. Unlike other marrow-derived cells and parenchymal cells that use the hypoxanthine-guanine phosphoribosyl transferase (salvage) pathway, activated lymphocytes rely predominantly on the de novo pathway for purine synthesis. This functional selectivity allows lymphocyte proliferation to be specifically targeted with less anticipated effect on erythropoiesis and neutrophil production than is seen with azathioprine.

Early studies in human kidney and heart transplant recipients showed that MMF, when substituted for azathioprine in standard triple-therapy regimens, is well tolerated and more efficacious than azathioprine. In a large, double-blind, randomized multicenter study comparing MMF versus azathioprine (with CyA and prednisone) involving 650 patients, the MMF group was associated with significant reduction in mortality as well as a reduction in the requirement for rejection treatment. However, there was noted to be an increase in the incidence of opportunistic viral infections in the MMF group. The overall greater efficacy of MMF compared to azathioprine has resulted in MMF generally replacing azathioprine in triple immunosuppressive protocols together with steroids and cyclosporine in most solid organ recipients.

Corticosteroids

Steroids are routinely used in almost all immunosuppressive protocols after organ transplantation. The metabolic side effects of steroids are well known and lead to significant morbidity and mortality in the post-transplant period. Almost 90% of organ recipients continue to receive prednisone at 1-year post-transplant and 70% at three-years post-transplant. A recent review of over 1800 patients from a combined registry outlined the morbid complications that patients suffer within the first year after transplantation. Many of these complications are known side effects of prednisone, including hypertension (16%), diabetes mellitus (16%), hyperlipidemia (26%), bone disease (5%) and cataracts (2%). It is thereby obvious that avoidance of steroids may decrease morbidity and mortality after organ transplantation. Two general approaches are used to institute prednisone-free immunosuppression: early and late withdrawal.

Withdrawal of prednisone during the first month post-transplant has resulted in long-term success of steroid withdrawal in 50–80% of patients. In these studies, the use of antilymphocyte antibody induction therapy appears to increase the likelihood of steroid withdrawal. Several centers have reported their results with immunosuppressive regimens that did not include steroids in the early post-transplant period. Studies reporting high success rates of 80% have used specific enrolment criteria, such as excluding patients with recurrent acute rejections or those with female gender. Review of numerous studies demonstrate that steroid free maintenance immunosuppression is possible in atleast 50% of patients, is as safe as triple drug therapy and may reduce some of the long-term complications of steroids. Owing to the fact that the majority of acute rejection episodes occur in the first three months post-transplant, steroid withdrawal is made after this time period, resulting in long-term success in about 80% of patients. Generally, there is no need for conventional induction agents when late withdrawal of steroids is done.

Anti-Lymphocyte Antibody Therapy

Despite the extensive use of induction therapy using anti-lymphocyte antibody in solid organ transplantation, their exact role is unclear.

There is no doubt that routine use of these agents is unwarranted as the generalized immunosuppression induced by then increased the risk of infections and malignancy. Despite the lack of consistent data supporting the routine use of induction therapy with antilymphocyte antibody agents, there is a role in certain select situations. Specifically, patients with early post-operative renal or hepatic dysfunction may benefit especially by the avoidance of cyclosporine therapy while using these induction agents. Antilymphocyte antibody therapy can provide effective immunosuppression for atleast 10 to 14 days without CyA or tacrolimus therapy. It has also been suggested that patients with overwhelming postoperative bacterial infections or diabetics with severe postoperative hyperglycemia may benefit from the comparatively low doses of corticosteroids required during antilymphocyte induction therapy.

The two main types of induction agents have been either the polyclonal antilymphocyte or antithymocyte globulins and more recently the murine monoclonal antibody OKT3. While these agents have been shown to be effective in terminating acute allograft rejection and in treating refractory rejection, the results of comparative studies of outcomes with and without monoclonal induction therapy have varied, with most studies demonstrating an effect on rejection that is maintained only while antibody therapy is ongoing. Without repeated administration, these agents only delay the time to a first rejection episode without decreasing the overall frequency or severity of rejection. More importantly, their use has been associated with an increased risk of short-term (infections) and long-term (lympho-proliferative disorders) complications. A complication specific to OKT3 is the development of a “flu-like syndrome” characterized by fever, chills and mild hypotension, typically seen with the first dose.

Since antilymphocyte antibodies are produced in nonhuman species, their use is associated with the phenomenon of sensitization, leading to decreased effectiveness with repeated use as well as the possibilty of serum sickness. The development of sensitization has been linked with an increased risk of acute vascular rejection. While this association has not been reported by other centers using OKT3 prophylaxis, it is believed that the development of immune-complex disease, inadequate immunosuppression due to decreased OKT3 levels or that OKT3 sensitization may be a marker for patients at higher risk for humoral rejection may be responsible for this phenomenon.

Interleukin-2 Receptor Inhibition

A new class of drugs has been developed which targets the high affinity IL-2 receptor. This receptor is present on nearly all activated T cells but not on resting T cells. In vivo activation of the high-affinity IL-2 receptor by IL-2 promoted the clonal expansion of the activated T cell population. A variety of rodent monoclonal antibodies directed against the á chain of the receptor have been used in animals and humans to achieve selective immunosuppression by targeting only T-cell clones responding to the allograft. Chimerisation or humanisation of these monoclonal antibodies resulted in antibodies with a predominantly human framework that retained the antigen specificity of the original rodent monoclonal antibodies. A fully humanized anti-IL2R monoclonal antibody, daclizumab, and a chimeric anti-IL-2R monoclonal antibody, basiliximab, have undergone successful phase III trials demonstrating their efficacy in the immunoprophylaxis of patients undergoing renal and cardiac transplantation.

Both agents have immunomodulatory effects that are similar to those of other monoclonal antibody-based therapies (i.e., induction of clonal anergy rather than clonal deletion). The advantages of these agents include their lack of immunogenicity, long half-lives, ability to repeat dosing, and short-term safety profile. Daclizumab appears to be an effective adjuvant immunomodulating agent in cardiac allograft recipients. It has advantages over conventional induction therapy as it is more selective and can be used for prolonged and potentially repeated periods. Studies with larger cohorts are needed to further study the short-term and long-term survival benefits for patients following organ transplantation and should determine the optimal dosing schedules of these new agents.

Stem Cell Transplantation and Immunosuppression

Materno-Fetal Tolerance

As outlined above, when tissues from an HLA-disparate donor are transplanted into a recipient they are always recognized as foreign, and immunosuppression is required to prevent rejection. An important exception to this is observed in pregnant women who tolerate their unborn fetus despite the fact that it expresses a full set of non-maternal HLA antigens inherited from the father. The mechanisms by which embryonic tissue demonstrates immune privilege during prenatal development have not yet been fully elucidated, however it is evident that interactions between fetus and mother differ substantially from the events triggered by a classical allograft. Consequently, much work is being dedicated to the emerging field of materno-fetal immunobiology in order to enable the development of innovative strategies to induce tolerance and prevent allogeneic graft rejection.

When maternal T cells encounter the fetus they demonstrate adaptive tolerance. In part this may be due to the absence of expression of MHC class II antigens and low levels of expression of MHC class I antigens on fetal cells. However, this can only partly explain the state of prolonged maternal tolerance since induction of HLA class I and II molecules inevitably occurs as the fetus matures and differentiates, yet rejection still does not occur. Consequently, non-fetal aspects of the placental barrier must be of critical importance in maintaining prolonged tolerance to the fetus. An important mechanism may relate to upregulation of the human nonclassical HLA class Ib antigen, designated HLA-G, by the syncytiotrophoblast. HLA-G molecules bind to inhibitory receptors on natural killer cells and subsequently protect against maternal rejection responses. The placenta produces high levels of the antiinflammatory cytokine interleukin 10 which stimulates HLA-G synthesis while concomitantly downregulating MHC class I antigen production, thus contributing to the tolerance-inducing local environment. The trophoblast also produces high levels of the enzyme indoleamine 2,3-dioxygenase, which catabolizes tryptophan, an essential amino acid necessary for rapid T cell proliferation.

Annexin II, found in isolated placental membranes in vitro is present in placental serum, exerts immunosuppressive properties, and additionally contributes to fetal allograft survival. Together, these features indicate that materno-fetal tolerance results from a combination of transiently reduced antigenicity of the fetus in combination with a complex tolerance-inducing milieu at the placental barrier.

Immunogenic Characteristics Oof Embryonic And Adult Stem Cells

Murine and human embryonic stem (ES) cells do not express HLA class I and II antigens, and demonstrate reduced surface expression of co-stimulatory molecules important for T cell activation.

Transplantation of murine ES cells demonstrates long-term graft survival despite the fact that these cells do acquire HLA class II antigen expression after in vivo differentiation. Since they are able to accomplish long-term engraftment without the need for immunosuppression, their inability to induce an immune response is not likely to be the result of escaping immune surveillance, but rather due to their ability to colonize the recipient thymus and induce intrathymic deletion of alloreactive recipient T cells.

Recently, a population of cells has been described in human adult bone marrow that has similar functional characteristics to embryonic stem cells in that they have high self-regenerating capability and capacity for differentiation into multiple cell types, including muscle, cartilage, fat, bone, and heart tissue. While such cells, termed adult mesenchymal stem cells (MSC), appear to have a more restricted self-renewal capacity and differentiation potential than ES cells, their functional characteristics may be sufficient for clinically meaningful tissue regeneration. A striking recent observation is that MSC can broadly inhibit T-cell proliferation and activation by various types of antigenic stimulation, including allogeneic stimuli. MSCs have been shown to inhibit both naive and memory T cell responses in a dosedependent fashion and affect cell proliferation, cytotoxicity, and the number of interferon gamma (IFN-gamma)-producing T cells. MSCs appear to inhibit T cell activation through direct contact, and do not require other regulatory cellular populations. Similarly to ES cells, adult bone marrow-derived mesenchymal stem cells (MSCs) do not express HLA class II molecules, and only low levels of HLA class I molecules. Despite the fact that MSC can be induced to express surface HLA class II molecules by in vitro culture with cytokines such as interferon-gamma, their ability to inhibit T cell activation results in induction of T cell non-responsiveness to the MSC themselves, endowing them with potential survival advantages in the setting of transplantation.

Tolerogenic Effects of Stem Cell Transplantation

Extending the approaches discussed above using donor-derived blood transfusions to induce a tolerogenic state to the subsequent organ, the most promising clinical strategy for tolerance induction at present is the use of donor-derived hematopoietic stem cells in conjunction with reduced myeloablative conditioning. The objective of this therapy is to achieve a state of so-called mixed chimerism, or the permanent co-existence of donor- and recipient-derived blood cells comprising all the different hematopoietic lineages in the same host. This approach has been tested in a variety of small and large animal settings and currently available data suggest that stable engraftment of donor bone marrow reliably renders the host tolerant to donor antigens and subsequently to any cellular or solid organ graft of the same donor.

The two underlying mechanisms by which creation of a mixedchimeric host results in tolerance induction are (1) thymic deletion of potentially donor-specific alloreactive T cells, and (2) nonthymic peripheral mechanisms, such as blocking costimulatory T cell activation, which facilitate the process of donor bone-marrow or stem cell engraftment. However, despite the efficacy of an approach using fully HLA-mismatched stem cells in an allogeneic host to induce tolerance to a subsequent organ allograft, the host is placed at a high risk of substantial morbidity and mortality due to toxicity of the myeloablative conditioning regimen and potential for graftversus- host disease, or immune-mediated attack of the host by the implanted allogeneic stem cells.

In an attempt to overcome these potential limiting toxicities, investigators have suggested the use of either adult bone marrowderived mesenchymal stem cells or preimplantation-derived embryonic stem (ES) cells for induction of mixed chimerism. The theoretical advantages of these cell types is their low level of surface expression of HLA class I and II antigens, and reduced surface expression of co-stimulatory molecules important for T cell activation. Rat preimplantation stage derived embryonic-like stem cells have been shown to successfully engraft in the recipient bone marrow without the need for pre-conditioning therapies such as irradiation, cytotoxic drug regimens or T cell depletion. Long-term partial mixed chimerism by use of rat preimplantation stage derived embryonic-like stem cells did not trigger graft-versus-host reactions, in contrast to the high frequency of this complication in the clinical setting of allogeneic hematopoietic stem cell transplantation. Of most interest, the induced partial chimerism enabled the recipient animals to be tolerant to a subsequent heart allograft. Allograft acceptance required the presence of an intact thymus, and rat ES cells were present in the recipient thymus.

Similar results have been reported following transplantation of human adult bone marrow-derived mesenchymal stem cells (MSC) into fetal sheep early in gestation, before and after the expected development of immunologic competence. In this xenogeneic system, human MSC engrafted, differentiated in a site-specific manner, and persisted in multiple tissues for as long as 13 months after transplantation, including the thymus. Since MSCs do not present alloantigen and do not require MHC expression to exert their inhibitory effect on alloimmune reactivity, the possibility exists that they could theoretically be derived from a donor irrespective of their HLA type and used to inhibit T-cell responses to transplantation antigens of an unrelated third party. In initial human clinical studies, the use human adult bone marrow-derived mesenchymal stem cells has been shown to successfully enable engraftment of subsequently infused allogeneic bone marrow in transplant recipients, reduce the risk of graft-versus-host disease, and reduce the need for concomitantly administered immunosuppression. Whether similar results will be obtained when combining adult bone marrow-derived mesenchymal stem cells with solid organ allografts remains to be determined, and this is an area of active research for clinical transplant immunobiologists. Of broader relevance, if the results relating to long-term engraftment and survival of adult bone marrowderived MSC are confirmed and extended in human clinical studies, they will have broad implications for the field of tissue and organ regeneration.

Recent Developments in Stem Cell Research and Therapy

Research using human and animal stem cells is an extremely active area of current biomedical inquiry. It is contributing new knowledge about the pathways of normal and abnormal cell differentiation and organismal development. It is opening vistas of new cell transplantation therapies for human diseases.

Although the availability of a variety of human stem cells is relatively recent—the isolation of human embryonic stem cells was first reported only in 1998—much is happening in both publicly funded and privately funded research centers around the world. It is difficult for anyone to stay abreast of all the results now rapidly accumulating.

To help us fulfill our mandate to "monitor stem cell research," the President's Council on Bioethics asked several experts to survey the recent published scientific literature and to contribute articles on various areas of stem cell research to this report (see articles by Drs. Gearhart, Ludwig and Thomson, Verfaillie, Prentice, Itescu, and Jaenisch in the Appendices). These reviews and the present chapter emphasize peer-reviewed, published work with human stem cells through July 2003. Interested readers should also consult the wide variety of other review articles that have appeared. This chapter should be read in conjunction with the commissioned review articles cited above. It draws on their findings, as well as on the Council's own monitoring activities, but it makes no attempt to summarize all the complexity of stem cell research or the vast array of results. Rather we offer here some general observations and specific examples that might help non-scientist readers understand the overall state of present human stem cell research, its therapeutic promise, and some of the problems that need to be solved if the research is to yield sound knowledge and clinical benefit. To that end, we highlight the importance of well-characterized, stable preparations of stem cells for obtaining reproducible experimental results, and we identify several problems that must be solved before these requirements can be fully met.

This study describes, by way of illustration and example, some of the better-characterized adult and embryonic stem cells. It also indicates some of the specific investigations that are being conducted with their aid. Finally, it considers how human stem cells are being used to explore their potential for treating disease, using experiments in animal models of Type- 1 diabetes as an example, and it points out some of the difficulties that must be overcome before stem cell-based remedies may be available to treat human diseases.

We confine our attention here to newly identified types of human stem cells and their potential use in research and future medical treatment. Accordingly, we do not consider those stem cell types that are already well established in medical practice and research. Specifically, we will not examine those preparations of bone marrow cells that have been clinically used for some years to treat various forms of anemia and cancer. Neither will we deal with hematopoietic (bloodforming) stem cells that have been isolated and purified from bone marrow and are now being intensively studied. Although these developments lie beyond the scope of this report, the demonstrated usefulness of these cells for research and therapy encourages many researchers to expect similar benefits from the newer stem cells that we shall consider here.

Cells and Their Derivatives

The adult human body, and all its differentiated cells, tissues, and organs, arise from a small group of cells contained within the early embryo at the blastocyst stage of its development. During in vivo embryonic development, these cells, constituting the inner cell mass (ICM), will divide and differentiate in concert with each other and with the whole of which they are a part, eventually producing the specialized and integrated tissues and organs of the body. But when embryos are grown [using in vitro fertilization (IVF)] in a laboratory setting, these ICM cells may be removed and isolated, and under appropriate conditions some will proliferate in vitro and become embryonic stem cell lines.

These embryonic stem cells are capable of becoming many different types of differentiated cells if stimulated to do so in vitro. However, it is not yet clear that the cells that survive the in vitro selection process to become embryonic stem cells have all of the same biological properties and potentials as the ICM cells of the blastocyst. In particular, it is not known for certain that human embryonic stem cells in vitro can give rise to all the different cell types of the adult body. As noted in the Introduction to this report, stem cells are a diverse class of cells, which can now be isolated from a variety of embryonic, fetal, and adult tissues. Stem cells share two characteristic properties: (1) unlimited or prolonged selfrenewal (that is, the capacity to maintain a pool of stem cells like themselves), and (2) potency for differentiation, the potential to produce *more differentiated cell types*—usually more than one *and, in some cases,* many. When stem cells head down the pathway toward differentiation, they usually proceed by first giving rise to a more specialized kind of stem cell (sometimes called "precursor cells" or "progenitor cells"), which can in turn either proliferate through self-renewal or produce fully specialized or differentiated cells.

At the top of the figure is an undifferentiated stem cell; in the central box are more "specialized" stem cells (or "precursor cells" or "progenitor cells"); at the bottom are various differentiated cells that are derived from the specialized stem cells. Dashed arrows indicate symmetrical (in the sense that both the daughter cells are stem cells) cell divisions that produce more stem cells (self-renewal). Solid arrows indicate asymmetric cell divisions that produce more differentiated daughter cells. (There may also be self-renewal with asymmetric division—not shown here—in which one daughter cell initiates a differentiation pathway while the other remains a stem cell.) Differentiation signals can be supplied by both soluble proteins and by specific, cell-surface binding sites.

Some of the specialized stem cells inside the dashed box, for example, mesenchymal stem cells, can be isolated from tissues after birth and correspond to adult stem cells. Scientists are currently investigat*ing wh*ether, at least in some cases, the process can be reversed, that is, whether specialized cells may, on appropriate signals, dedifferentiate to become precursor or even fully und*ifferentiated st*em cells.

The terminology used to describe different stem cell types can be confusing. As used in this chapter, stem cells are selfrenewing, cultured cells, grown an*d preserved* in vitro, that are capable—upon exposure to appropriate signals—of differentiating themselves into (usually more than one) specialized cell types. Stem cells may be classified either according to their origins or according to their developmental potential.

Stem cells may be obtained *from various sources: f*rom embryos, from fetal tissues, from umbilical cord blood, and from tissues of adults (or children). Thus, depending on their origin, stem cell preparations may be called adult st*em cel*ls, embryonic stem cells, embryonic germ cells, or fetal stem cells.

Adult stem cells are cells derived from various tissues or organs in humans or animals that have the two characteristic properties of stem cells (self-renewal and potency for differentiation). Embryonic stem cells (ESCs) are derived from cells isolated from the inner cell mass of early embryos. Embryonic germ cells (EGCs) are stem cells.

derived from the primordial germ cells of a fetus. Fetal stem cells (not further discussed in this chapter, but included for the sake of completeness) are derived from the developing tissues and organs of fetuses; because they come (unlike EGCs) from already differentiated tissues, they are (like adult stem cells) "non-embryonic," and may be expected to behave as such.

Depending on their developmental potential, cells may be called pluripotent, multipotent, or unipotent. Cells that can produce all the cell types of the developing body, such as the ICM cells of the blastocyst, are said to be pluripotent. The somewhat more specialized stem cells, of the sort found in the developed organs or tissues of the body, are said to be multipotent if they produce more than one differentiated tissue cell type, and unipotent if they produce only one differentiated tissue cell type.

We introduce in this chapter an additional term: stem cell preparation. A stem cell preparation is a population of stem cells, prepared, grown, and preserved under certain conditions. Because different laboratories (or even the same one) can have different preparations of the same type of stem cell, it is important to recognize the potential differences between particular preparations of embryonic stem cells. It will sometimes be important to call attention to this fact, by speaking of a "preparation of ES cells" (or a preparation of adult stem cells) rather than of "ES cells," pure and simple.

We will use the term "stem cell preparations" when we are speaking of a diverse group of stem cell cultures, when we are speaking of stem cell cultures that contain an admixture of other types of cells, or when the developmental homogeneity of the stem cells in the population has not been defined.

Adult and embryonic stem cell populations have also been called "stem cell lines." In the past, the term "cell line" denoted a cell population (usually of cancer cells containing abnormal chromosome numbers or structure, or both) that could grow "indefinitely" in vitro. Embryonic and some adult stem cell preparations are capable of prolonged growth beyond 50 population doublings in vitro while retaining their characteristic stem cell properties and initially with no change in the chromosome numbers and structure. It is not yet known whether any preparation of human ES cells (generally believed to be much longer-lived than adult stem cells) will continue to grow "indefinitely," without undergoing genetic changes.

Under the influence of various cell-differentiation signals, embryonic stem cells differentiate into numerous distinct types of more specialized cells. Some of these are specialized stem cells that can also self-renew, while retaining their ability also to differentiate into multiple cell types. Recent research has led to the isolation of an increasing number of adult (non-embryonic) stem cells (dashed box area of Figure 1) from such tissues as bone marrow (for example, hematopoietic and mesenchymal stem cells), brain (for example, neural stem cells) and other tissues. Although these stem cell preparations differ from one another in their future fates, they tend to be grouped together (especially in the public policy debates) under the name "adult stem cells," even though they may have been obtained from children or even from umbilical cord blood obtained at the time of childbirth.

Subsequent exposure to additional differentiation signals can cause these specialized stem cells to differentiate further, so that they finally give rise to the variety of differentiated cells that make up the adult body (labeled A-D in Figure 1). At each stage of the differentiation process, specific sets of genes are expressed (or "turned on") and other sets are repressed (or "turned off"), to produce the specific proteins that give each cell its distinctive properties. At each stage along the way, proteins called transcription factors play key roles in determining which sets of genes are expressed and repressed, and therefore what sort of a cell the newly differentiated cell will become.

Reproducible Results Using Stem Cell Preparations and THeir Derivatives

A major goal of scientific research is the acquisition of reliable knowledge based on experiments that yield reproducible results. Reproducible results are possible only if the materials used in experiments remain constant and stable.

To obtain reproducible results in experiments using stem cells, it is essential to produce, preserve, characterize, and continually re-characterize preparations of stem cells in ways that increase the likelihood that the cells used to repeat experiments will remain unchanged—a technically challenging task. The tendency of stem cells in vitro to differentiate spontaneously into more specialized cells makes the task of obtaining homogeneous and stable stem cell preparations especially challenging, and much basic research is needed to learn how to control the fate of these cells. Failure to control the cells may yield experimental results that are difficult or impossible to reproduce. The following more specific observations make clear the dimensions of this difficulty.

A. Initial Stem Cell Preparations: Can Contain Multiple Cell Types Isolation of adult stem cells from source tissues such as bone marrow, brain, or muscle initially yields a heterogeneous cell preparation. The initial preparation contains the several cell types found in the source tissue, and it may also include red blood cells, white blood cells, and (possibly) circulating stem cells, owing to the presence of blood flowing through the tissue in question. Initial mixtures of cells may then be treated in various ways to remove unwanted contaminating cells, thereby increasing the proportion of stem cells in the preparation. But seldom, if ever, does one produce an adult stem cell preparation that is 100 percent stem cells, unless the adult stem cell preparation has been "single-cell cloned" in vitro.

The way in which human embryonic stem cells have been produced from ICM cells also raises a question about the "species homogeneity" of the initial cell preparations. In the past, human embryonic stem cells were isolated and maintained by *in vitro* growth on top of irradiated (so that they no longer divide) "feeder layers" of mouse cells. It is thought that the feeder cells secrete factor(s) that enable the stem cells to divide while maintaining a relatively undifferentiated state.

Although the mouse cells have been treated to prevent their cell division, should any of them happen to survive, human embryonic stem cells prepared in this way may contain some viable mouse cells. More recently, several groups have shown that it is possible to grow ESCs on feeder layers of human cells, including fibroblasts obtained from skin biopsies, or without any feeder cell layer at all. One way to be certain that human embryonic stem cell preparations do not contain any mouse feeder cells is through "single cell cloning".

B. Genetically Homogenous Stem Cells through Single Cell Cloning: Some preparations of stem cells growing in vitro have been "single cell cloned," that is, grown as a population derived from a single stem cell. By placing a cylinder over a single *cell l*ocated with a microscope, scientists are able to isolate within the cylinder all the progeny produced by subsequent cell divisions beginning from this single cell. The result is a stem cell preparation in which all the cells are descended from the original single cell. The cells within the cylinder are then harvested and grown to greater numbers in vitro, and the resulting stem cell preparation is said to bc "single cell cloned." The stem cells within a "single cell cloned" population are, at least to begin with, genetically homogeneous because they are all derived *from the sam*e original cell. Some of the ESC preparations produced prior to August 9, 2001 have been "single cell cloned." * The issue of possible mouse virus contamination is dealt with in Section F, below.

C. Expansion in Vitro, Preservation, and Storage : Reproducible results require that preparations of stem cells, even if genetically homogenous when first isolated, remain stable over time and during preservation. This, too, is not a simple matter with stem cells, despite the fact that the selfrenewal characteristic of human embryonic and adult stem cells enables them—unlike differentiated cells from many human tissues—to be grown in large numbers in vitro while maintaining their essential stem cell characteristics. After such expansion, many, presumably identical, vials of the cells can be frozen and preserved at very low temperatures. Frozen stem cell preparations can later be thawed and grown again in vitro to produce larger numbers of cells.

As with all dividing cells, stem cells are subject to a very small but definite chance of mutation during DNA replication; thus, prolonged growth in vitro could introduce genetic *heterogeneity into an originally homogeneous population. During this process of repeated expansion and* preservation, subtle changes in the growth conditions or other variables may give rise to "selective pressures" that can increase the heterogeneity in a stem cell preparation by favoring the multiplication of advantaged cell variants in the population. It is not known at present how many of the 78 human ESC preparations, designated as eligible for federal funding under the current policy, have developed genetic variants that may make them unsuitable for further research.

Whether several cycles of freezing and thawing change the phenotypic characteristics of stem cell preparations needs detailed study. However, the practical advantages of preserving stem cell preparations by freezing are too large to ignore. Such preservation makes it possible to repeat an experiment many times with a very similar stem cell preparation. It would also make it possible, should stem cell based therapies be developed in the future, to treat multiple patients with a common, well-characterized cell preparation derived from a single initial stem cell sample.

D: Chromosome Changes: In addition to the possible loss of homogeneity in stem cell preparations owing to variability in growth conditions or to freezing and thawing, there is the possibility of variation being introduced during the processes of growth and cell division.

Normal human stem cells (like all human somatic cells) have 46 chromosomes. During the copying of chromosomal DNA and the separation of daughter chromosomes at cell division, rare mistakes occur that lead to the formation of abnormal chromosomes or maldistribution of normal ones. Cells with abnormal chromosomes or chromosome numbers can progress to malignancy, so retention of the normal human chromosome number and structure is an essential characteristic of useful human stem cell preparations. The most studied preparations of human stem cells generally have normal human chromosome numbers and structure. Nevertheless, vigilance is needed, for even a small number of chromosomally abnormal cells could end up causing cancer in future clinical trials of stem cell based therapies.

E. Developmental Heterogeneity of Stem Cell Preparations: The in vitro growth conditions and the presence of specific chemicals or proteins, or both, in the culture medium can influence the differentiation pathway taken by stem cells as they start to differentiate. Thus, even initially homogeneous, "single cell cloned" stem cell preparations may become developmentally heterogeneous over time, with respect *to the percenta*ge of cells in the preparation that are in one or another differentiated state. For example, a stem cell preparation after growth in vitro under specific conditions might contain 75 percent fully differentiated (insulin-producing) cells and 25 percent partially differentiated cells. The biological properties of the fully differentiated cells and the partially differentiated cells are likely to be different. If such a cell preparation is used in research, or transplanted into an animal model of human disease and a biological effect is observed, one must do additional experiments to determine whether the effect was due to the fully differentiated cells or to the partially differentiated cells (or perhaps to both acting together) in the now mixed preparation.

F. Microbial Contamination: Stem cell preparations originally isolated from humans and expanded in vitro may also be variably contaminated with human viruses, bacteria, fungi, and mycoplasma. ESC preparations isolated using mouse feeder cell layers might also be contaminated with mouse viruses. Specific tests need to be performed on the source tissue and periodically on the resulting stem cell preparations to rule out the presence of these contaminants. Some of these contaminants can also multiply when stem cells are grown in vitro, and their presence can influence the results obtained when stem cell preparations are used in subsequent experiments. The presence of such contaminants can also potentially affect the reproducibility of the results of experiments in which stem cell preparations are studied in vivo in experimental animals.

In summary, there are numerous challenges to obtaining and preserving the uniform and stable preparations of stem cells necessary for reliable research and, eventually, for safe and effective possible therapies. Researchers must address multiple factors in order to maximize the probability of obtaining reproducible results with human stem cell preparations. Human stem cell preparations that are "single cell cloned," with a normal chromosome structure and number, and • stored as multiple samples that are preserved at very low temperature, and compared in experiments where cells from the same lot of frozen material are used, and well-characterized as to the absence of cellular, viral, bacterial, fungal, and mycoplasma contaminants, and tested to determine the proportion of stem cells *and* various differentiated cells in the cell preparation used in the experiments, are most likely to yield experimental results that will be reproducible. Preparations with these properties will be the most useful both in basic research and in investigations of possible clinical applications.

III. Major Examples of Human Stel Cells

In this section we discuss major examples of human

stem cells that meet many of the criteria listed above. Among human adult stem cells, we focus on mesenchymal stem cells *(MSCs)*, multipotent adult progenitor cells (MAPCs), and neural stem cells, and among human embryonic stem cells, on ESC and EGC cells. For infor*mation on* the wide variety of other human stem cell preparations isolated from adult tissues, see reference (4) (Appendix K).

Further research on some of these other adult stem cell preparations may demonstrate that they can also be "single cell cloned," expanded considerably by growth in vitro with retention of normal chromosome structure and number, and preserved by freezing and storage at low temperatures. At that point, it would be very important to compare the properties of these other adult stem cells, and the more differentiated cells that can be derived from them, with the already characterized human embryonic and adult stem cell preparations.

Human Adult Stem Cells

Human Mesenchymal Stem Cells

Bone marrow contains *at least two major kinds of stem* cells: hematopoietic stem cells, which give rise to the red cells and white cells of the blood, and mesenchymal stem cells, which can be reproducibly isolated and expanded in vitro and that can differentiate in vitro into cells with properties of cartilage, bone, adipose (fat), and muscle cells. The characteristics (morphology, expressed proteins, and biological properties) of these cells have been somewhat difficult to specify, because they appear to vary depending upon the in vitro culture conditions and the specific cell preparation. However, there is a recent report indicating that MSCs, if isolated using three somewhat different methods, give rise to stem cell preparations whose properties are very similar to one another. Using dual antibody staining and fluorescence-activated cell sorting, Gronthos and colleagues isolated human MSCs in almost pure form and expanded them substantially in vitro. Thus, human MSC preparations isolated in different laboratories by different methods may have similar but not identical properties.

A molecular analysis of genes expressed in a single-cellderived colony of MSCs provided evidence for the activity of genes also turned on in bone, cartilage, adipose, muscle, hematopoiesis-supporting stromal, endothelial, and neuronal cells. These results are surprising in that MSCs derived from a single cell appear to be expressing genes associated with multiple major cell lineages. It is possible that different cells within the colony had already entered into distinct differentiation pathways, resulting in a developmentally heterogeneous population composed of *several different* cell types.

Mesenchymal stem cells are important for research and therapy for several reasons. First, because they can be differentiated in vitro into multiple cell types, they make possible detailed research on the molecular events underlying differentiation into bone, cartilage, and fat cell lineages.

Second, they have recently been shown to support the in vitro growth of human embryonic stem cells. Thus, they could replace the mo*use feeder* cells used previously, obviating the need to satisfy FDA requirements for xenotransplantation, should the ESCs or their derivatives ever be used in human clinical research or transplantation therapy. Third, clinical studies are already underway in which MSCs are cotransplanted with autologous hematopoietic stem cells into cancer patients to replace their blood cell-forming system, destroyed by radiation or high dose chemotherapy. It is believed that the MSCs will support the repopulation of the bone marrow by the injected hematopoietic stem cells.

In addition, injecting allogeneic MSCs (MSCs from a genetically different human donor) may also prove valuable in modulating the immune system to make it more accepting of foreign tissue grafts. Finally, MSCs have the potential for cell-replacement therapies in injuries involving bone, tendon, or cartilage and possibly other diseases. They are, in fact, already being tested as experimental therapies for osteogenesis imperfecta, metachromatic leukodystrophy, and Hurler syndrome. These last two studies are of great interest, since allogeneic MSCs were used and no serious adverse immune reactions were noted.

Multipotent Adult Progenitor Cells (MAPCs).

Verfaillie and coworkers recently described the isolation of MAPCs from rat, mouse, and human bone marrow. Like MSCs, MAPCs can also be differentiated in vitro into cells with the properties of cartilage, bone, adipose, and muscle cells. In addition, there is evidence for the in vitro differentiation of human MAPCs into functional, hepatocyte-like cells, a potential that has not so far been shown for MSCs. There is increasing interest in MAPCs, both as potential precursors of multiple differentiated tissues and, ultimately, for possible autologous transplantation therapy.

The relationship between human MSCs and the human MAPCs described by Verfaillie and coworkers needs to be clarified by further research. Both kinds of cells are isolated from bone marrow aspirates as cells that adhere to plastic. Each can be differentiated in vitro into cells with cartilage, bone, and fat cell properties. They express several of the same cell antigens, but are reported to differ in a few others. MAPCs have to be maintained at specific, low cell densities when grown in vitro, otherwise they tend to differentiate into MSCs. It remains important that the isolation and properties of MAPCs be reproduced in additional laboratories.

Human Neural Stem Cells.

The nervous system is made up of three major types of cells, neurons or nerve cells proper, and two kinds of supporting or glial cells (oligodendrocyte, astrocyte). Stem cells capable of differentiating into one or more of these neural cell lineages can be isolated from brain tissue (particularly the olfactory bulb and lining of the ventricles) and grown in vitro. In the presence of purified growth-factor proteins, the population of cells can be expanded by growth in vitro as round clumps of cells called neurospheres. However, many neurospheres grown in culture are developmentally heterogeneous in that they contain more than one neural cell type, and the number of self-renewing cells is frequently low (less than five percent). Although neural stem cells are still insufficiently understood, they are already proving valuable in basic research on neural development. The ability to grow reproducible neural stem cells in vitro has facilitated identification of important neural stem cell growth factors and their cellular receptors.

For example, human neural stem cells from the developing human brain cortex, expanded in culture in the presence of leukemia inhibitory factor (LIF), allowed growth of a selfrenewing neural stem cell preparation for up to 110 population doublings. Withdrawal of LIF led to decreased expression of about 200 genes, which were specifically identified through use of "gene chips" manufactured by Affymetrix. These genes are presumably involved in promoting or preserving the stem cell's capacity for self-renewal in the undifferentiated state.

The number and specificity of the molecular changes characterized in these experiments powerfully illustrate the usefulness of neural and other stem cell preparations in basic biomedical research.

Human neural stem cells are also being injected into animals to test their effects on animal models of human neurological disease. To track the fate of the introduced human cells, they must first be modified or "marked" in ways that permit their specific detection. Marked human neural stem cells are easily tracked after they are injected into experimental animals, making it possible to determine whether they survive and migrate following injection. Studies of this type have provided evidence that human neural cells can migrate extensively in the brain after injection. In addition, such cells can be injected into animal models of human diseases such as intracerebral hemorrhage and Parkinson Disease (PD) to study their effect on the progression of the disease. Although human neural stem cells may not yet be as well characterized as MSCs or ESCs, they are being actively studied with the hope that they can be used in future treatments for devastating neurological diseases such as Alzheimer Disease and PD.

Adult Stem Cells from Other Sources.

Prentice has *summarized a large amount of recent inf*ormation on preparations of stem cells isolated from amniotic fluid, peripheral blood, umbilical cord blood, umbilical cord, brain tissue, muscle, liver, pancreas, cornea, salivary gland, skin, tendon, heart, cartilage, thymus, dental pulp, and adipose tissue. Studies of many of the stem cell preparations from these sources are just getting started, and further work is needed to determine their biological properties and their relatedness to other stem cell types. In some cases, the long-term expandability in vitro of these stem cells has not been demonstrated. Yet, the demonstration that they can be isolated from such tissue compartments in animals should spur the search for similar human stem cell types.

As Prentice also reports, many attempts have already been made using various preparations of adult stem cells to influence or alter the course of diseases in animal models.

Despite the fact that the stem cell preparations used are not well characterized, and reproducible results have yet to be obtained, preliminary findings are sometimes encouraging. It is of course not yet clear whether the injected cells are functioning as stem cells, fusing with existing host cells, or stimulating the influx of the host's own stem cells into the target tissue. But, if reproduced, these preliminary findings may point the way to future therapies, even in the absence of precise knowledge of the mechanism(s) of cellular action.

Human Embryonic Stem Cells 1. Human Embryonic Stem Cells (ESCs).

Human embryonic *stem cells have been isolated from the* inner cell masses of blastocyst-stàge human embryos in multiple laboratories around the world. There is great interest in understanding the properties of these cells because they hold out the promise of being able to be differentiated into a large number of different cell types for possible cell therapies, as contrasted with the more limited number of cell types available by differentiation of specific adult stem cell preparations. As of July 2003, 12 ESC preparations (up from 2 such preparations a year earlier) out of a total of 78 "eligible" preparations of human ESCs were available for shipment to recipients of U.S. federal research grants. The review by Ludwig and Thomson lists more than 40 peer-reviewed human ESC primary research papers that have been published since the initial publication in 1998.

Although isolated from different blastocyst-stage human embryos in laboratories in different parts of the world, ESCs have a number of properties in common. These include the presence of common cell surface antigens (recognized by binding of specific antibodies), expression of the enzymes alkaline phosphatase and telomerase, and production of a common gene-regulating transcription factor known as Oct-4. At least 12 different preparations of ESCs have been expanded by growth in vitro, frozen and stored at low temperature, and at least partially characterized. Some of these ESC preparations have been "single-cell cloned." Human ESCs have been differentiated in vitro into neural (neurons, astrocytes, and oligodendrocytes), cardiac (synchronously contracting cardiomyocytes), endothelial (blood vessels), hematopoietic (multiple blood cell lineages), hepatocyte (liver cell), and trophoblast (placenta) lineages. In the case of neural and cardiac lineages, similar results have been obtained in different laboratories using different preparations of ESCs, thus fulfilling the "reproducible results" criterion described above. For other lineages, the results described have not yet been reproduced in another laboratory.

Embryonic Germ Cells

Human embryonic germ cells *are isolated from the primordial* germ tissues of aborted fetuses. Gearhart has summarized the results of recent research with human and mouse EG cells. One study focused on regulation of imprinted genes in EG cells: it showed "that general dysregulation of imprinted genes will not be a barrier to their (EG cell) use in transplantation studies." In addition, Kerr and coworkers showed that cells derived from human EG cells, when introduced into the cerebrospinal fluid of rats, became extensively distributed over the length of the spinal cord and expressed markers of various nerve cell types. Rats paralyzed by virus-induced nerve-cell loss recovered partial motor function after transplantation with the human cells. The authors suggested that this could be due to the secretion of transforming growth factor-á and brain-derived growth factor by the transplanted cells and subsequent enhancement of rat neuron survival and function.

Until recently, work with human EG cells came primarily from one laboratory. Recently the isolation and properties of human EG cells have been independently confirmed. Because human EG cells share many (but not all) properties with ESCs, these cells offer another important avenue of inquiry.

Embryonic Stem Cells from Cloned Embryos (Cloned ESCs).

Although it has yet to be accomplished in practice, somatic cell nuclear transfer (SCNT) could create cloned human embryos from which embryonic stem cells could be isolated that would be genetically virtually identical to the person who donated the nucleus for SCNT: hence cloned ESCs. In theory, using such cloned embryonic *stem cells* from individual patients might provide a way around possible immune rejection, though in practice this could require individual cloned embryos for each prospective patient—a daunting task. And clinical uses might require a separate FDA approval for every single cloned stem cell line or its derivatives.

The ability to produce cloned mouse stem cells and genetically modify them in vitro has made possible an experiment demonstrating the potential of cloned human embryonic stem cells in the possible future *treatment* of human genetic diseases. Rideout et al. used a mutant *mouse* strain that was deficient in immune system function. They produced a cloned mouse embryonic stem cell line carrying the mutation, and *then specifically* repaired that gene mutation in vitro. The repaired cloned stem cell preparation was then differentiated in vitro into bone marrow precursor cells. When these precursor cells were injected back into the genetically mutant mice, they produced partial restoration of immune system function.

Production of cloned human embryonic stem cell preparations remains technic*ally* very difficult and ethically controversial. Recently however, Chen and coworkers have reported that fusion of human fibroblasts with enucleated rabbit oocytes in vitro leads to the development of embryolike structures from which cell preparations with properties similar to human embryonic stem cells can be isolated. This work needs to be confirmed by repetition in other laboratories.

In addition, further work is needed to decisively settle the question of whether rabbit (or human egg donor) mitochondrial DNA and rabbit (or human egg donor) mitochondrial proteins persist in the embryonic stem cell preparations. Persistence of these foreign mitochondrial proteins in these human ESClike preparations could possibly increase the probability of immune rejection of the cloned cells, thus limiting their clinical application, although the immune reaction might not be as severe as that to foreign proteins produced under the direction of chromosomal genes. The presence of foreign or aberrant mitochondria also carries the risk of transmitting mitochondrial disease (caused by defects in mitochondrial DNA) that could be detrimental to the cells and to the recipient into whom they might eventually be transplanted.

Basic Research Using Human Stem Cells

Human stem cells are proving useful in basic research in several ways. They are useful in unraveling the complex molecular pathways governing human differentiation. For example, because ESCs can be stimulated in vitro to produce more differentiated cells, this transition can be studied in greater detail and under better-controlled conditions than it can be in vivo. In the best circumstances, these differentiated cells can be grown as largely homogeneous cell populations, and their gene expression profiles can be compared in detail.

Also, stem cell preparations can be used to produce populations of specialized cells that are not easily obtained in other ways. In one case, for example, this approach has provided large quantities of human trophoblast-like cells that have not been previously available. In addition, cultures of differentiated cells derived from stem cells could be used to test new drugs and chemical compounds for toxicity and mutagenicity. As experience with these differentiated derivatives of human ESCs grows, it may become possible to reduce or eliminate the use of live animals in such testing protocols.

In the near future, the differentiated state of various human cell types will be characterized not just by a few biological markers, but by the pattern and levels of expression of hundreds or thousands of genes. Integration of this knowledge with the catalog of all human genes produced during the Human Genome Project will gradually give us knowledge of which genes are key regulators of human development and which genes are central to maintaining the stem cell state. Increased understanding of the molecular pathways of human cell differentiation should eventually lead to the ability to direct in vitro differentiation along pathways that yield cells useful in medical treatment. In addition, when the normal range of gene expression patterns is known, researchers can then determine which genes are expressed abnormally in various diseases, thus increasing our understanding of and ability to treat these diseases.

Human Stem Cells and the Treatment of Disease

A major goal of stem cell research is to provide healthy differentiated cells that, once transplanted, could repair or replace a patient's diseased or destroyed tissues. In pursuit of this goal, one likely approach would start by isolating stem cells that could be expanded substantially in vitro. A large number of the cultivated stem cells could then be stored in the frozen state, extensively tested for safety and efficacy as outlined above, and used as reproducible starting material from which to prepare differentiated cell preparations that will express the needed beneficial properties when they are transplanted into patients with specific diseases or deficiencies.

To make more concrete both the potential of this approach and the obstacles it faces, we will summarize, as a case study example, some current information on the properties of cells derived from human stem cell populations that have been used in an animal model of Type-1 diabetes. But before doing so, we discuss an obstacle to any successful program of stem cellbased transplantation therapy: the problem of immune rejection of the transplanted cells.

Will Stem Cell-Based Therapies Be Limited by Immune Rejection?

Much of the impetus for human stem cell research comes from the hope that stem cells (or, more likely, differentiated cells derived from them) will one day prove useful in cell transplantation therapies for a variety of human diseases. Such cell transplantation would augment the current practice of whole organ transplantation. To the extent that the healing process works with in vitro derived cells, the need for organ donors and long waiting lists for organ donation might be reduced or even eliminated.

Will the recipient (patient) accept or reject the transplanted human cells? In principle, the problem might seem avoidable altogether: adult stem cells could be obtained from each individual patient needing treatment. They could then be grown or modified to produce the desired (autologous and hence rejection-proof) transplantable cells. But the logistical difficulties in processing separate and unique materials for each patient suggest that this approach may not be practical.

The cost and time required to produce sufficient numbers of well-characterized cells suitable for therapy suggest that it will be cells derived from one or another unique stem cell line that will be used to treat many (genetically different) individual patients (allogeneic cell transplantation).

When allogeneic organ or tissue transplantation is currently done using, for example, bone marrow, kidney, or heart, powerful immunosuppressive drugs—carrying undesirable side effects—must be used to prevent immunological rejection of the transplanted tissue. Without such immunosuppression, the patient's T-lymphocytes and natural killer (NK) cells recognize surface molecules on the transplanted cells as "foreign" and attack and destroy the cells. Also, in whole organ transplantation, donor T-lymphocytes and NK cells, entering the recipient with the transplanted organ, can also destroy the tissues of the transplant recipient (called "graft versus host" disease).

Are the differentiated derivatives of human stem cells as likely to incite immune rejection, when transplanted, as are solid organs? Do their surfaces carry those protein antigens that will be recognized as "foreign"? Experiments have been done to examine human ESC and MSC preparations growing in vitro for the expression of surface molecules known to play important roles in the immune rejection process. Drukker and coworkers showed that embryonic stem cells in vitro express very low levels of the immunologically crucial major histocompatibility complex class I (MHC-I) proteins on their cell surface. The presence of MHC-I proteins increased moderately when the ESCs became differentiated, whether in vitro or in vivo. A more pronounced increase in MHC-I antigen expression was observed when the ESCs were exposed to gamma-interferon, a protein produced in the body during immune reactions. Thus, under some circumstances, human ESC-derived cells can express cell surface molecules that could lead to immune rejection upon allogeneic transplantation.

Similarly, Majumdar and colleagues showed that human mesenchymal stem cells in vitro express multiple proteins on their cell surfaces that would enable them to bind to, and interact with, T-lymphocytes. They also observed that gammainterferon increased expression of both human leukocyte antigen (HLA) class I and class II molecules on the surface of these MSCs. These results indicate that it will probably not be possible to predict, solely on the basis of in vitro experiments, the likelihood that transplanted allogeneic MSCs would trigger immune rejection processes in vivo.

Many further studies in this area are badly needed. At this time there is insufficient information to determine which, if any, of the approaches to get around the rejection problem will eventually prove successful.

Case Study: Stem Cells in the Future Treatment of Type- 1 Diabetes?

1. The Disease and Its Causes.

The human body converts the sugar glucose into cell energy for heart and brain functioning, and indeed, for all bodily and mental activities. Glucose is derived from dietary carbohydrates, is stored as glycogen in the liver, and is released again when needed into the bloodstream. A protein hormone called insulin, produced by the beta cells in the islets of the pancreas, facilitates the entrance of glucose from the bloodstream into the cells, where it is then metabolized. Insulin is critical for regulating the body's use of glucose and the glucose concentration in the circulating blood.

The body's failure to produce sufficient amounts of insulin results in diabetes, an extremely common metabolic disease affecting over 10 million Americans, often with widespread and devastating consequences. In some five to ten percent of cases, known as Type-1 diabetes (or "juvenile diabetes"), the disease is caused by "autoimmunity," a process in which the body's immune system attacks "self." T-lymphocytes attack the patient's own insulin-producing beta cells in the pancreas.

Eventually, this results in destruction of ninety percent or so of the beta cells, resulting in the diabetic state.

With a deficiency or absence of insulin, the blood glucose becomes elevated and may lead to diabetic coma, a fatal condition if untreated. Chronic diabetes, both Type-1 and the much more common Type-2 diabetes (which is not autoimmune, but largely genetic), causes late complications in the retina, kidneys, nerves, and blood vessels. It is the leading cause of blindness, kidney failure, and amputations in the U.S. and a major cause of strokes and heart attacks.

Type-1 diabetes is a devastating, lifelong condition that currently affects an estimated 550,000-1,100,000 Americans, including many children. It imposes a significant burden on the U.S. healthcare system and the economy as a whole, over and above the disabilities and impairments borne by individual sufferers. Recent estimates suggest that treatment of all forms of diabetes costs Americans a total of $132 billion per year. At 5-10 percent of all diabetes cases, the costs of Type-1 diabetes can be estimated as $6.5-$13 billion per year.

2. Current Therapy Choices and Outcomes.

The current *treatment of Type-1 diabetes consists of* insulin injections, given several times a day in response to repeatedly measured blood glucose levels. Although this treatment is lifeprolonging, the procedures are painful and burdensome, and in many cases they do not adequately control blood glucose concentrations. Whole pancreas transplants can essentially cure Type-1 diabetes, but fewer than 2,000 donor pancreases become available for transplantation in the U.S. each year, and they are primarily used to treat patients who also need a kidney transplant. Like all recipients of donated organs, pancreas transplant recipients must continuously take powerful drugs to suppress the immunological rejection of the transplanted pancreas.

In addition to treatment with whole pancreas transplantation, small numbers of Type-1 diabetes patients have been treated by transplantation of donor pancreatic islets into the liver of the patient coupled with a less intensive immunosuppressive treatment (the Edmonton protocol). Expanded clinical trials of this procedure are currently underway. Scientists are also evaluating methods of slowing the original autoimmune destruction of pancreatic beta cells that produces the disease in the first place.

Whole pancreas and islet cell transplants ameliorate Type- 1 diabetes, but there is nowhere near enough of these materials to treat all in need. To overcome this shortage, people hope that human stem cells can be induced—at will and in bulk— to differentiate in vitro into functional pancreatic beta cells, available for transplantation. Of course, it would still be crucial to prevent immunological destruction of the newly transplanted stem cell-derived beta cells.

3. Stem Cell Therapy for Type-1 Diabetes?

Initial exp*eriments in mice suggested that*

*insulinpro*ducing cells could be obtained from mouse embryonic stem cells following in vitro differentiation. Can this approach be extended to human stem cells? A number of attempts have been made, with promising initial findings, yet they are not easily evaluated, partly because the criteria for characterizing the cells are not standardized. In a recent paper, Lechner and Habener provided a list of six criteria to define the characteristics of pancreas-derived "beta-like" cells that could be potentially useful in treatment of Type-1 diabetes. We have used those criteria to facilitate assessment of the current state of progress toward development of functional "beta-like" cells that might eventually be tested in Type-1 diabetes patients. Table 1 summarizes and compares the properties of human cell preparations recently produced in research seeking this objective by Abraham et al., Zulewski et al., Assady et al., Zhao et al., and Zalzman et al., and tested in mouse models of human diabetes.

As the results described in Table 1 indicate, cells derived from some human stem cells transplanted into specific strains of mice mimicking major aspects of Type-1 human diabetes were able to reverse high blood glucose concentrations.

Although these results are encouraging, the transplant rejection question remains unanswered because the likely immune rejection of the transplanted human cells was prevented in these experiments by using special strains of immunodeficient mice that lack the capacity to recognize and attack foreign cells.

No tumors were observed in the transplanted mice, but the experiments were terminated after about three months, an insufficient time for much tumor development to occur. Because many Type-1 diabetes patients are children and because a largely effective therapy (insulin injection) is currently available, the introduction of islet cell transplant therapy will need a high degree of certainty that the introduced cells or their derivatives will not become malignant over the course of the patient's life. Stringent tests of the cancer-causing potential of candidate cell preparations will be required, including multiyear studies in animals that live longer than mice or rats. Longterm follow-up of children and adult patients who had received bone marrow transplants many years ago has revealed an increased risk of severe neurologic complications and a variety of types of cancer. C. Therapeutic Applications of Mesenchymal Stem Cells (MSCs) Before stem cell based therapies are used to treat human diseases, they will have to gain approval through the Food and Drug Administration (FDA) regulatory process. The first step in this process is filing an Investigational New Drug (IND) application. As of July 2003, four IND applications have been filed for clinical applications of mesenchymal stem cells.

The disease indications include: (1) providing MSC support for peripheral blood stem cell transplantation in cancer treatment, (2) providing MSC support for cord blood transplantation in cancer treatment, (3) using MSCs to stimulate regeneration of cardiac tissue after acute myocardial infarction (heart attack), and (4) using MSCs to stimulate regeneration of cardiac tissue in cases of congestive heart failure. The first two applications are currently in Phase II of the regulatory process, with pivotal Phase III trials scheduled to begin in 2004. D. Evaluating the Different Types of Stem Cells A major unresolved issue at present involves the therapeutic potential of human adult stem cells compared with embryonic stem cells. The answer may well be different for different diseases and for patients of different ages. For example, in treating an elderly patient with Parkinson's Disease, the use of adult stem cells may be appropriate even if these cells may have a more limited number of cell divisions remaining. On the other hand, treating a child with Type I Diabetes, one may want to use embryonic stem cells because of their potentially greater longevity, or other factors. The only valid way to resolve these questions is by instituting rigorous therapeutic trials which test the efficacy of the different types of stem cells in treating a variety of different diseases to determine their comparative efficacy. Clearly, such trials would be a long-term endeavor, since it would take years to obtain answers to these very critical questions.

Private Sector Activity

In the United States, much of the basic research on animal stem cells and human adult stem cells has been publicly funded. Yet before *2001,* research in the U.S., using human ESCs could only be done in the private sector (the locus also of much research on animal and human adult stem cells). The current state of knowledge about human ESCs (and also about human MSCs) reflects pioneering and on-going stem cell research funded by the private sector in the U.S. For example, the work that led to the 1998 reports of the first isolation of both ESCs and EGCs, was funded by Geron Corporation. Embryonic and adult stem cell research is today vigorously pursued by many companies and supported by several private philanthropic foundations, and the results of some of this research have been published in peer-reviewed journals. Private sector organizations have pursued and been awarded patents on the stem cells themselves and methods for producing and using them to treat disease. As noted above, at least one company (Osiris Therapeutics) has protocols under review at the FDA for clinical trials with MSCs. It seems likely that private sector companies will continue to play large roles in the future development of stem cell based therapies.

Preliminary Conclusions

While it might be argued that it is too soon to attempt to draw any conclusions about the state of a field that is changing as rapidly as stem cell research, we draw the following preliminary conclusions regarding the current state of the field.

Human stem cells can be reproducibly isolated from a variety of embryonic, fetal, and adult tissue sources. Some human stem cell preparations (for example, human ESCs, EGCs, MSCs, and MAPCs) can be reproducibly expanded to substantially larger cell numbers in vitro, the cells can be stored frozen and recovered, and they can be characterized and compared by a variety of techniques. These cells are receiving a large share of the attention regarding possible future (nonhematopoietic) stem cell transplantation therapies.

Preparations of ESCs, EGCs, MSCs, and MAPCs can be induced to differentiate in vitro into a variety of cells with properties similar to those found in differentiated tissues.

Research using these human stem cell preparations holds promise for: (a) increased understanding of the basic molecular process underlying cell differentiation, (b) increased understanding of the early stages of genetic diseases (and possibly cancer), and (c) future cell transplantation therapies for human diseases.

The case study of developing stem cell-based therapies for Type-1 diabetes illustrates that, although insulin-producing cells have been derived from human stem cell preparations, we could still have a long way to go before stem cell-based therapies can be developed and made available for this disease. This appears to be true irrespective of whether one starts from human embryonic stem cells or from human adult stem cells. The transplant rejection problem remains a major obstacle, but only one among many.

Human mesenchymal stem cells are currently being evaluated in pre-clinical studies and clinical trials for several specific human diseases.

Much basic and applied research remains to be done if human stem cells are to achieve their promise in regenerative medicine. This research is expensive and technically challenging, and requires scientists willing to take a long perspective in order to discover, through painstaking research, which combinations of techniques could turn out to be successful. Strong financial support, public and private, will be indispensable to achieving success.

Bibliography

Alan Meisel of the University of Pittsburgh School of Law, quoted in Coyle, M., "The Clone Zone," *The National Law Journal*, May 15, 2002.

Brivanlou, A. H., *et al.,* "Stem cells. Setting standards for human embryonic stem cells," Science 300: 913-916 (2003).

Bush, G.W., "Stem Cell Science and the Preservation of Life," *The New York Times*, August 12, 2001, p. D13.

Callahan, D., *What Price Better Health: The Hazards of the Research Imperative.*

Clark, J., "Squandering Our Technological Future," *The New York Times*, August 30, 2001, p. A19. See also the presentation of Thomas Okarma, President and CEO of Geron Corporation, before the Council on September 4, 2003.

Conference of Catholic Bishops, in a news release ("Catholic Bishops Criticize Bush Policy on Embryo Research," made available by the United States Conference of Catholic Bishops, August 9, 2001).

Englund, U., *et al.,* "Transplantation of human neural progenitor cells into the neonatal rat brain: extensive migration and differentiation with longdistance axonal projections," Experimental Neurology 173: 1-21 (2002).

Faden, *et al.*, "Public Stem Cell Banks: Considerations of Justice in Stem Cell Research and Therapy" *Hastings Center Report* 33:6 (2003).

Faraci, M., *et al.*, "Severe neurologic complications after hematopoietic stem cell transplantation in children," Neurology 59: 1895-1904 (2002).

Kerr, D. A., *et al.*, "Human Embryonic Germ Cell Derivatives Facilitate Motor Recovery of Rats with Diffuse Motor Neuron Injury," The Journal of Neuroscience 23: 5131-5140 (2003).

Latham, S., "Ethics and Politics," *American Journal of Bioethics* 2(1): 46 (2002).

Lumelsky, N., *et al.*, "Differentiation of embryonic stem cells to insulin-secreting structures similar to pancreatic islets," Science 292: 1389-1394 (2001).

Majumdar, M. K., *et al.*, "Characterization and functionality of cell surface molecules on human mesenchymal stem cells," Journal of Biomedical Science 10: 228-241 (2003).

Marquis, D., "Stem cell research: The Failure of Bioethics," *Free Inquiry* 22(1): Winter 2002.

Murray, T., "Hard Cell," *The American Prospect,* September 24, 2001.

Okarma, T., Presentation at the September 4, 2003, meeting of the President's Council on Bioethics, Washington, D.C., available at www.bioethics.gov.

One useful description of this approach to the issue is Zoloth, L., "Jordan's Banks, A View from the First Years of Human Embryonic Stem Cell Research," *American Journal of Bioethics* 2(1): 7 (2002).

Onyango, P., *et al.*, "Monoallelic expression and methylation of imprinted genes in human and mouse embryonic germ cell lineages," Proceedings of the National Academy of Sciences of the *United States of America* 99: 10599-10604 (2002).

Orr, R., "The Moral Status of the Embryonal Stem Cell: Inherent or Imputed?" *American Journal of Bioethics* 2(1): 57-59 (2002).

Sato, N., *et al.*, "Molecular signature of human embryonic stem cells and its comparison with the mouse," Developmental Biology 260: 404-413 (2003).

Soria, B., *et al.*, "Insulin-secreting cells derived from embryonic stem cells normalize glycemia in streptozotocin-induced diabetic mice," Diabetes 49: 157-162 (2000)

Xu, R. H., *et al.*, "BMP4 initiates human embryonic cell differentiation to trophoblast," Nature Biotechnology 20: 1261-1264 (2002).

Zalzman, M., *et al.*, "Reversal of hyperglycemia in mice using human expandable insulin-producing cells differentiated from fetal liver cells," Proceedings of the National Academy of Sciences of the *United States of America 100: 7253-7258 (2003).*

Zhao, M., et al., "Amelioration of streptozotocin-induced diabetes in mice using human islet cells derived from long-term culture in vitro," Transplantation 73: 1454- 1460 (2002).

Index

N

O

P

R

S

T

U

V

Z